CHAINING OREGON

SURVEYING THE PUBLIC LANDS OF THE PACIFIC NORTHWEST, 1851–1855

Surveyors are not heroic figures. They come after the explorers, they douse with system what was once the incandescent excitement of danger and the unknown. They conquer nothing but ignorance, and if they are surveying a boundary they are so bound by astronomical and geodetic compulsions that they might as well run on rails. The mythic light in which we have bathed our frontier times, when decision was for the individual will and a man tested himself against wild beasts, wild men, wild weathers, and so knew himself a man — that light does not shine on them as it shines on trapper and trader and scout and cowboy and Indian fighter. They do not even acquire the more pedestrian glamour of the farming pioneer, though they make him possible.[1]

WALLACE STEGNER

CHAINING OREGON

SURVEYING THE PUBLIC LANDS OF THE PACIFIC NORTHWEST, 1851–1855

by

Kay Atwood

The McDonald & Woodward Publishing Company
Blacksburg, Virginia

The McDonald & Woodward Publishing Company

Blacksburg, Virginia, and Granville, Ohio

CHAINING OREGON: SURVEYING THE PUBLIC LANDS OF THE PACIFIC
NORTHWEST, 1851–1855

© 2008 by Kay Atwood

Printed in the United States of America
by McNaughton & Gunn, Inc., Saline, Michigan

First printing July 2008

15 14 13 12 11 10 09 08   10 9 8 7 6 5 4 3 2 1

Library of Congress Cataloging-in-Publication Data

Atwood, Kay.
  Chaining Oregon : surveying the public lands of the Pacific Northwest,
    1851–1855 / by Kay Atwood.
    p. cm.
  Includes bibliographical references and index.
  ISBN 978-0-939923-20-5 (pbk. : alk. paper)
  1. Northwest, Pacific—Discovery and exploration. 2. Northwest, Pacific—
  Geography. 3. Northwest, Pacific—Surveys—History—19th century.
  4. Surveyors—Northwest, Pacific—History—19th century. 5. Land
  settlement—Northwest, Pacific—History—19th century. 6. Frontier and
  pioneer life—Northwest, Pacific. I. Title.
  F852.A89 2008
  917.95′04302—dc22

                                                            2008019267

Contents

For

David Stoddard Atwood

Foreword

Reading Kay Atwood's book, *Chaining Oregon*, is a rewarding experience. The author engagingly relates the monumental task undertaken by the first government land surveyors in America's Pacific Northwest — "running lines" and "setting corners" across the barely settled valleys and steep forested mountains of what is today western Oregon and western Washington. She ably places this challenging enterprise, one that remains little known to the general public and to most regional historians, into its broader historical context. One example: How much the nation's increasingly bitter politics during the early 1850s — Whig versus Democrat, North versus South — disrupted life in the Far West, including the frustrating delays and legal confusion caused by political patronage decisions back in the nation's capital that determined the appointment of surveyors in distant Oregon.

The survey crews' difficulty in negotiating Oregon's often-rugged terrain, with its deep forests and thick underbrush, was compounded by the Northwest's wet-season weather. Rain made such work not merely miserable but technically difficult. To read their township survey notes for the months between October and April is to wonder how the crew chiefs ever kept their crucial notebooks' pages from dissolving to mush in the rain. The fast-paced years from 1850 through

1855 presented many other challenges, such as when the federal surveyors, for the most part educated Yankees hailing from quiet small towns, found themselves quite literally in the midst of a gold-rush frenzy and a bitter Indian war in the Rogue River Valley.

Chaining Oregon. The title is, of course, to be taken both literally and metaphorically: the surveyors' then-standard tool of horizontal measurement, the 66-foot-long "Gunther's chain" made of 100 metal links, as well as the conceptual capturing and subduing — the "chaining" — of a heretofore "wild" land by the superimposition of the Enlightenment's Cartesian grid. The surveyors' township and range lines, their far more numerous section lines, exactingly and exhaustingly measured and marked, did not merely determine the Northwest's present-day land-ownership patterns and the wanderings of many of its roads with their right-angle turns, seemingly located without rhyme or reason. Little-known men like James Freeman, Butler Ives, and George Hyde first brought the Pacific Northwest into the rationalized, computed, and ultimately commodified world of today.

The township plats (maps) and accompanying descriptive field notes of Oregon's early-day surveyors have recently come to be recognized as a treasure trove of environmental data. Historians, ecologists, wildlife biologists, and others now mine them for what can be gleaned about the Northwest's landscape a century and a half ago — before the land had become *enchained* by fence and field, by city blocks and suburban lots. My hunch is that Atwood's work too will find a ready audience among the region's historians, as well as its still-busy ranks of land surveyors who, by necessity, tend to become well-versed historians of their own craft. But general readers living in the Northwest, especially those with a historical curiosity about their own patch of ground, will also find the story enlightening and well-told.

In closing, it's also been my own pleasure to read this book, in part because the story it tells brought back memories of my first day of work with the US Forest Service — long before I went on to a career with the agency as an academically trained historian and

archaeologist. Having, through sheer dumb luck, obtained a temporary job as a lowly chainman on the Rogue River National Forest's
survey crew, I was informed that first morning that I was to assist the
Forest's experienced and physically tireless cadastral surveyor, Tom
Newcomb. In hindsight, I've come to wonder if this assignment was a
test to see how well I did in the woods, trying to keep up with him as
Tom led me on a grueling trek packing gear up and down poison-oak-
infested, 50-percent slopes in southwestern Oregon's Applegate Valley. We were "chasing a corner," and for me, at the time barely knowing the difference between an azimuth and an alidade, his running
narrative about setting corners, scribing bearing trees, taking solar
shots, and repeated mention of the seemingly hallowed names "Ives
and Hyde" was almost like hearing a foreign language. *Chaining Oregon* helps it all make sense.

JEFF LALANDE

Preface

In 1972, I stood in the Jackson County Surveyor's Office in Medford, Oregon, looking at 1854 General Land Office maps of the Bear Creek Valley. My project at the time — a comparative study of the valley landscape in the early 1850s and the 1970s — began here. "By whom surveyed?" posed a question at the bottom of the maps and on each page appeared the answer: Ives and Hyde. "Butler Ives and George Hyde," then County Surveyor Mark Boyden said, pulling down volumes that accompanied the maps. He helped me navigate the field notes, explaining that the government surveyors re-copied their original notes and sketches for office draftsmen to use in creating the final township maps. Surveyor and son of a surveyor, Mark had spent years tracking the field work of Ives and Hyde and he knew well the value of their precise calculations. "The best of the best," he called Butler Ives.

I used those records many times through the decades and wondered about the surveyors who spent time here so many years ago. Mark was always willing to talk with me about "Ives and Hyde" and generously shared his own research. As my interest in Pacific Northwest history deepened, my curiosity about these men grew. Who were they and how did they contribute toward settlement? I began my research ten years ago and between my professional obligations and sometimes instead of them, the story of the surveyors took shape.

A number of important books contributed to my background study. Wallace Stegner's *Beyond the Hundredth Meridian* and Richard A. Bartlett's *Great Surveys of the American West* discuss important nineteenth century surveying expeditions in the West. Authors who treat the history of the public land surveys under the direction of the US General Land Office include Lowell O. Stewart, *Public Land Surveys* (1935; reprinted 1979); Francois D. Uzes, *Chaining the Land: A History of Surveying in California* (1977); Lola Cazier, *Surveys and Surveyors of the Public Domain 1785–1975* (1977); and C. Albert White, *A History of the Rectangular Survey System* (1983). Collectively, these works provide a comprehensive overview of public land surveying history in the United States.

Other books on my reading list addressed the historical, geographical, sociological, and environmental aspects of settlement in western Oregon, including *Empire of the Columbia: A History of the Pacific Northwest* by Dorothy O. Johansen and Charles M. Gates (1967); *The Willamette Valley: Migration and Settlement on the Oregon Frontier* by William A. Bowen (1978); *Environment and Experience: Settlement Culture in Nineteenth-Century Oregon* by Peter G. Boag (1992); and *Landscapes of Promise: The Oregon Story 1800–1940* by William G. Robbins (1997). Andro Linklater's *Measuring America: How an Untamed Wilderness Shaped the United States and Fulfilled the Promise of Democracy* (2002) offers a thoughtful, readable assessment of the importance of surveying in our history. The detailed, beautiful maps in the *Atlas of Oregon*, by Stuart Allan, Aileen R. Buckley, and James E. Meacham and edited by William G. Loy (2001), proved invaluable in determining the location of the survey contracts. Focused on broader historical themes, these authors all acknowledge the importance of the public land surveys to Oregon settlement but do not discuss the deputy surveyors or their experiences in any detail.

Published materials on the early government surveyors in Oregon were scarce. The recollections of Kimball Webster, a young government surveyor who worked in Oregon from 1851 to 1854, were published as *The Gold Seekers of '49: A Personal Narrative of the Overland Trial and Adventures in California and Oregon from 1849 to 1854* (1917). Informative articles in *The Oregon Surveyor*, a publication of the Professional Land Surveyors of Oregon, included Lane J. Bouman's study on donation land claims, William W. and Jeanne E. Glenn's biographical sketches of surveyors, and C. Albert White's study of the Willamette Meridian and Base Line surveys.

Three primary sources were especially important in understanding the technical and administrative aspects of the government surveys. One was *Elements of Surveying and Navigation* by Charles Davies, (1830 and later editions), a basic text used by Oregon's early surveyors. The other two, which deputies carried in the field, were the US General Land Office's *Instructions to Surveyor of Public Lands in Oregon* (1851) and *Instructions to the Surveyors General of Public Lands* (1855).

Government documents and manuscript collections yielded particularly valuable resources for research. The National Archives and Records Administration in Seattle houses US General Land Office field notes and plat maps of the Pacific Northwest, deputy surveyors' letters, and papers of the surveyors general that include correspondence with settlers, political associates, and superiors in Washington, DC. The Jackson County, Oregon, Surveyor's Office retains photocopies of the original field books of Ives and Hyde.

Original diaries and correspondence of Oregon's deputy surveyors survive in public and private collections in Ohio, Michigan, Minnesota, Illinois, California, and Oregon. The US government encouraged the surveyors to keep personal diaries to support their field notes and, although they tended toward utility rather than reflection, these volumes recorded camp locations, supplies acquired, leisure activities, as well as information about personnel. The diaries consistently detail two matters most critical to a contract's success: weather and miles surveyed each day. Quotations from these original journals are reproduced in *Chaining Oregon* as written, with all spelling and grammatical errors intact.

I am grateful to Western Reserve Historical Society for permitting me to quote from the Diaries of Butler Ives (MSS 3011) held in their collection. The Stanford University Libraries allowed me to use portions of Butler Ives's letter of February 4, 1853, held in the Papers of M. A. Ives, Timothy Hopkins Transportation Collection 1916–1942, Department of Special Collections and University Archives. The Detroit Public Library gave permission to use quotations from William Ives' Correspondence and Papers (MSS 1125) held in the Burton Historical Collection. John Preston's letter to George Hyde, July 14, 1855, is contained in the Louis Hyde Family Book, Reed Hyde Papers, a private collection. Dr. Stan Larson of the manuscripts division, J. Willard Marriott Library, University of Utah, granted permission to quote from "History Comes to the Plains," by Wallace Stegner. Martha Ives Duncan provided me with the text of Butler Ives' letter to William Ives, December 2, 1852. George Keith McFall and his daughter, Ellen McFall, assented to my quoting George McFall's Journal.

I thank the many people who helped me with my research for this book: Kathleen Leles DiGiovanni of the Oakland Public Library; John R. Gonzalez of the California State Library; Margaret Kimball of Special Collections, Stanford University Libraries; Sandy McGuire of the Clackamas County Historical Society; John Ferrell and Valoise Armstrong, National Archives and Records Administration, Seattle; Susan Seyl and Lucy Berkley, Oregon Historical Society; Laurie Klein of the Beinecke Rare Book and Manuscript Library, Yale University; Roger K. Gambrel of the Joliet Public Library; Michelle Sweetser of the Bentley Historical Library; Janet Bloom of the William L. Clements Library, University of Michigan; Mark E. Harvey of the State Archives of Michigan; Eric Moody of the Nevada Historical Society; Frances TeCulver of the Fenton History Center Museum and Library, Jamestown, New York; Nancy G. Karmazin, Grosse Ile Historical Society, Grosse Ile, Michigan; and Debra Moore, Archivist, Head, Acquisition and Special Media, Hudson's Bay Company Archives, Archives of Manitoba.

Margaret L. Steneck, Residential College, University of Michigan, guided my research on the Lodi Plains Academy southwest of Ann Arbor, Michigan. Anne Richards, Southern Oregon University Library and Julie Drengson of the Jackson County Library System, Medford, Oregon, tracked down obscure research materials. Maggie Weaver of the Bureau of Land Management Office, Portland, Oregon guided my search for General Land Office maps.

Many individuals offered information and encouragement. I thank Jack Bowder, Mary Elizabeth Braun, Philip Dole, Gail E. H. Evans-Hatch, Coni Florey, Mary Freymiller, David D. Hunting, Jr., Allen I. Hunting, Jr., John Hunting, Larry Hyder, Marianne Keddington-Lang, Dan Linscheid, Donna Logan, Marshall Lango, Richard Straw, Dennis Todd, and W. Thomas White.

Norman Caldwell, Martha Ives Duncan, Douglas H. Duncan, Harold Otness, Elaine Bowe Johnson, Jeff LaLande, and Roger Roberts read draft manuscripts whole or in part and offered thoughtful suggestions. Tom Newcomb of the Jackson County, Oregon, Surveyor's Office kindly shared his files with me; Roger Roberts, Jackson County Surveyor, answered my questions and made research material available. Stuart Allan of Allan Cartography, Medford, Oregon advised me regarding map design and he and his staff contributed their considerable artistic skills to produce maps for the book. Martha Ives Duncan, great-grandniece of William and Butler Ives, gave this project her unwavering support. Jeff LaLande graciously agreed to write the foreword for *Chaining Oregon*. I thank Mark E. Boyden for sharing his considerable knowledge of surveying history.

Nancy Parker applied her sharp eye and meticulous attention to detail to ready the manuscript for publication. I am grateful to Jerry McDonald of McDonald and Woodward for his interest in my manuscript and for guiding me through the publication process. Patricia Newcomb, McDonald and Woodward marketing manager, devoted considerable skill and hard work to introduce *Chaining Oregon* to the larger world. As always, I am deeply grateful to my husband, David, for his steadfast love and encouragement.

Introduction

Published histories illuminate the explorers, miners, and settlers who shaped Oregon, but leave the surveyors in shadow. Who were the men Stegner says fell outside the circle of light that washed over the "trapper and trader and scout and cowboy and Indian fighter"? Where did they come from, how long did they stay, and what did they accomplish?

Chaining Oregon: Surveying the Public Lands of the Pacific Northwest, 1851–1855 tells the story of the indefatigable, courageous men who first staked the Oregon Country, the area encompassing western Oregon and western Washington. It visits the surveyors' familial and geographic origins, investigates the seasons of their field work, the urban and rural landscapes they encountered, and the financial and political pressures that bore down on them. The first comprehensive history of the Pacific Northwest's early surveyors, *Chaining Oregon* uses the engineers' own observations recorded in field notes, journals, and correspondence to depict their personal experiences and hardships. *Chaining Oregon* shines some of Stegner's "mythic light" on James Freeman, William Ives, Butler Ives, George Hyde, and Joseph Hunt as they achieved work of immense scope while enduring great physical challenge.

These men were not lured to Oregon by a longing for gold or a lust for adventure. Trained as engineers and scientists, and gifted with

1

an artist's powers of observation, they came to do a job. In mid-March, 1851, William Ives and James Freeman, coming from the forests of Michigan and Wisconsin, risked the long sea voyage from the Atlantic Coast expressly to work as surveyors. According to Wallace Stegner, "Surveyors are not heroic figures. They douse with system what was once the incandescent excitement of danger and the unknown." In our mind's eye, we picture surveyors as deliberate men hunched over their tripods. We see them this way because mile after mile they did just that, alone save for a handful of assistants. The surveyors' isolation and the complexity of their occupation have kept us from fully recognizing the heroic qualities that enabled them to persevere and the magnitude of their contribution to Oregon's settlement.

As surveyors, Ives and Freeman and their colleagues on this enterprise, participated in a profession that had long been important in American settlement. George Washington began his career as a surveyor in 1748, at the age of sixteen, when he was invited to accompany a party surveying lands west of the Blue Ridge Mountains. Washington received appointment in 1749 to his first public office as surveyor of newly created Culpeper County, Virginia. In 1763, as a member of the company organized to drain Dismal Swamp in southeastern Virginia and northeastern North Carolina, Washington learned tenacity and meticulousness as he surveyed that area of dense forests and tangled undergrowth.

The rectangular survey system Freeman and Ives implemented in the Pacific Northwest grew out of the formal beginnings of the federal system established by Congress in the Land Ordinance of May 20, 1785. Prior to that time, colonial Anglo-America's rural population determined land boundaries by "metes and bounds," borrowing England's traditional definition of a parcel of land made "by citing the owners of abutting lands and describing the length of each course of a boundary as 'along' some apparent line, such as 'along a stream' or 'along the road.'"[1] In order to avoid the irregular and often confused surveys of the colonial era, the Land Ordinance of 1785 introduced an orderly system of laying one-mile-square parcels on federal lands. In a format used ever since, the Ordinance stipulated that surveyors establish a north-south principal meridian and an east-west base. The intersecting point of these two lines was known as the "initial" point. After identifying the meridian and the base, they were to calculate parallel lines of longitude and latitude six miles apart. In this way the surveyors created townships of 36 square miles. This acreage was further divided into one-mile squares of 640 acres.[2] To manage the government surveys, Congress created the Office of Surveyor General within the Treasury Department in 1796 and opened the General Land Office in 1812, charging the agency with supervising the public land surveys and handling sales and collections of monies resulting from the disposal of public lands to private ownership. On March 3, 1849, Congress created the Department of the Interior and placed the General Land Office under its jurisdiction.[3]

Two years later, when Ives and Freeman arrived in Oregon, their surveying practice adhered to this established measuring system. At the same time, it accelerated the expansion of settlement fostered by the Northwest Ordinance of 1787 and perpetuated by Americans' incessant hunger for land. From the 1790s to the 1840s, this Ordinance spread thousands of settlers across what is now considered the Midwest. Beyond the Midwest, however, lay the vast Great Plains and Rocky Mountains — a "Great American Desert" — a concept introduced by Zebulon M. Pike and perpetuated for decades in reports of the 1817–1820 expeditions of Colonel Stephen H. Long of the United States Topographic Engineers. Pike's report characterized the Great Plains and adjacent Rocky Mountain region as mostly arid, barren, and not generally suitable for agriculture.[4]

Leaping beyond the "Great American Desert," in the 1840s, America's new generation headed for Oregon, continuing the expansion of the nation's western boundary. Economic and industrial expansion in the 1840s prompted prospective settlers to look to the Pacific Northwest, seeing there, as historian Dorothy Johansen observes, the promise of free land "in a region they believed to have almost unlimited agricultural and commercial potentialities."[5] As the

Oregon-bound settlers sustained the pattern of westward movement, so the surveyors that accompanied them directly extended the linear survey checkerboard already etched across the Mississippi and Ohio valleys. These Midwest engineers, as they surveyed in both places, constituted a human link that joined the regions.

Initially, things went well for the surveyors in Oregon, but soon the realities of weather and politics set in. Endowed with strength and stamina as well as mathematical skill and scientific knowledge, Ives, Freeman and the others hauled cumbersome, delicate equipment mile after tortuous mile. They were forced to record their calculations while knee-deep in icy water or braced against chill wind on high peaks. Unlike explorers who could usually circumvent geographic obstacles in their path, the surveyors were obligated by their lines to cross whatever terrain lay ahead. In contrast to the adventurers who traveled the Oregon Country before Euro-American settlement began, the surveyors knew the frustrations of working while farmers interrupted them with questions or while vandals pulled out their hard-won stakes behind them. Not only physical impediments, but also political exigencies compounded their hardship. Toiling within the framework of the federal bureaucracy, the surveyors endured political manipulation. Although they were, as Stegner noted, "utilitarian only, [and] policy was none of their business," nevertheless they came near foundering more than once in the political squalls that battered the surveyor general's office.[6]

Between 1851 and 1855, these sinewy, weathered men — James Freeman, William Ives, Butler Ives, George Hyde, and Joseph Hunt — boosted the Territory's population by enabling the settlement of donation land claims and helping to establish commerce, both essential precursors to the stability and increased confidence that led to Oregon's statehood in 1859.

But their work did far more than make it possible to organize land ownership. In the long term, these men helped the Oregon Territory become a true part of the nation. In short, these surveyors helped to sustain the continuum of western expansion, shape the mid-nineteenth-century landscape, and promote growth. In doing so, each of them — at once scientist, engineer, and artist — left a detailed record of what the country looked like to their eye.

Whether we seek tangible information or our imaginations merely long to know, the surveyors gave us a unique way to see the past. Their maps — the first detailed diagrams of the Oregon Country — and field notes supplement the emigrants' letters that described their surroundings to loved ones back home, the journals of diarists who recorded the routines of daily life for their own reflection, and the visual images captured by painters and photographers. Objective in ways many artists and writers avoided, these meticulous, observant engineers scrutinized the landscape's broad canvas as well as its most minute detail. They catalogued landforms, wooded areas, open prairies, Indian villages, trails, roads, ferry crossings, and towns, precisely recording the points where these features intersected their lines. Settlers' claims, houses, barns, and cultivated fields all went into the record along with vegetation types and their extent, terrain characteristics and soil qualities.

Measuring with compass and chain between proscribed points on each township and section line, the surveyors faithfully logged the landmarks that intersected their path. As they took bearings to plot the direction of outlying objects with respect to the compass points, the men charted these features as well. Oregon's government surveyors stood linked to the heavens by their instruments while they inched forward on earth. "[T]he pole-star," their manual of instructions read, "very nearly reaches the true meridian when it is in the same vertical plane with the star Alioth in the tail of the Great Bear."[7] The surveyors' language springs from art and science. With the artist's eye for detail and the scientist's precision, they advanced settlement as significantly as did the farmers, merchants, and politicians. It is through their painstaking calculations and keen observations of the landscape they helped transform, that we come to know the men and their legacy.

Chapter 1 – Spring, 1851

Out to Oregon

" High Character and Integrity."

On May 3, 1851, the *New Orleans* steamed north along the California coast towards San Francisco Bay. Brothers William Ives and Butler Ives pressed hard against the deck rail. Weeks of confinement in the ship's cramped quarters left them yearning to stride again on solid ground. Butler Ives scanned the rolling landscape of the starboard side, the hills' timbered ridges and grass-covered slopes so close that he could make out small herds of cattle grazing in the sun.[1]

The *New Orleans* steamed through the Golden Gate into the crowded San Francisco harbor, its waters black with birds. On the boat's approach, seals slid silently from bare rocks into the sea. The crew dropped anchor near the landing and, after the customs officer had examined their luggage, the Ives brothers climbed into small boats to go ashore.[2] They dragged their trunks up the wharf's steep stairs into the crowded street where buildings stood like cranes on pilings above the tide. Turned down repeatedly in their search for lodging, it was sundown before the men found rooms and supper at the Jones Hotel.[3] Exhausted and relieved that the hard, dangerous passage was over, they slept.

Shouts and flickering light awakened William Ives at midnight. Outside his window, fire — "one of the most grand and terrific I ever

7

saw" — lit the sky over Clay Street. A west wind whipped the flames toward the bay, shifted and swept them back into the heart of town. When the blaze neared their hotel, the brothers dragged their trunks up the California Street hill where they watched through the night as the city dissolved in the "May 4" conflagration, San Francisco's sixth and worst fire in less than two years. As William and Butler Ives watched, the "brick & iron buildings did not withstand the fire atall but withered before it like dew."[4]

At dawn sixteen entire blocks and parts of several others lay in smoldering ruins. When the Ives brothers went out onto the charred streets later in the day to find the post office, they passed dazed residents searching in the rubble for salvage and setting up tents outside the burned area. Unlike these survivors or the merchants and prospectors they passed, William and Butler Ives faced another 650 miles at sea. They were US government surveyors and they were on their way to Oregon.

Oregon stretched over vast country west of the Rocky Mountains. The region's population clustered almost entirely in valleys west of the Cascades where, in the early 1830s, former Hudson's Bay Company (HBC) trappers and their families had settled French Prairie in the Willamette Valley and on the Tualatin Plains. North of the Columbia River, other former HBC retirees farmed the Cowlitz River Valley. American trappers and representatives of the US government intermittently visited the region, which Britain still considered under its control, while in the late 1830s protestant missionaries and Roman Catholic priests established more permanent settlements. After 1840, a small but steady stream of emigrants came West, and in 1846, when Britain and the US signed a treaty setting the 49th parallel as a boundary between their interests, more settlers made the Oregon Country their destination. Over 5,000 persons crossed overland the following year. While the majority of new arrivals chose the Willamette Valley, a number of Americans moved north of the Columbia taking land along the lower Cowlitz River in the vicinity of Monticello, the first county seat for Lewis County. The US Congress organized the Territory of Oregon as a temporary government in August, 1848, and the estimated six thousand emigrants who arrived in 1850 raised the population to almost 12,000 persons (Figure 1).[5]

Congress passed the Oregon Donation Land Act on September 27, 1850. Crafted to promote settlement in Oregon, the legislation had three key provisions: the selection of a surveyor general, the initiation of public land surveys west of the Cascade Range, and the award of land claims to settlers who met specific requirements. Through this Act, the government offered 320 acres to any white male over eighteen years of age who was a citizen or had declared his intention to apply, who resided in the Territory on or before December 1, 1850, and who had cultivated his fields for at least four years. A settler's wife, should he be married, could claim 320 acres in her own name.[6] Single men who settled in Oregon between December 1, 1850, and December 1, 1853, could claim 160 acres and a similar amount of land for their wives.

Reputedly, the first federal legislation to grant women property in their own right, the Act accommodated the very real possibility that a woman might be widowed en route to Oregon, or at least before the couple became fully established, thus preserving her right to retain title to part of the land. The provision also guaranteed an influx of women into the new Territory. Allowing men to double the size of their acreage if they were married set the stage for countless hasty or arranged marriages, some of these between adult men and child brides.

After a claimant submitted "notification" — proof of settlement, cultivation, and citizenship — and after the US government had surveyed the land, the General Land Office issued the settler a certificate. In the spring of 1851, thousands of people lived south of the Columbia River on lands they measured in terms of trees, streams, and hills. People who had more ground than they could farm wanted to divide and sell. Needing their land measured to do so, they pressed for the surveys and official designation of their claims' boundaries.

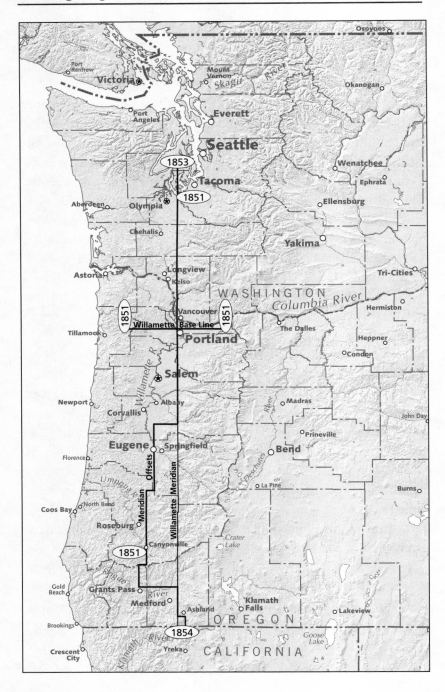

One day after the Donation Act passed Congress, President Millard Fillmore appointed William Gooding, Illinois-Michigan Canal trustee and former canal superintendent, as Oregon's first surveyor general. Gooding declined the offer. On November 22, 1850, during the Congressional recess, Fillmore named John B. Preston, Gooding's colleague and a former Illinois-Michigan Canal engineer, to the position. Preston accepted; and, on December 16, 1850, the president submitted the nominee's name for Senate approval (Figure 2).[7]

Appointments were more often than not politically inspired, and many new surveyors general understood little about land surveying or budgets. Thirty-three-year-old John Bower Preston excelled at both. Born in Granville, Vermont, on May 11, 1817, he spent his early childhood along the Metawee River, relocating when his father moved the family to Warsaw, New York, and later to Rochester. In search of richer farmland, the Prestons moved west again in 1836, settling in Will County, Illinois.[8] Young Preston studied bookkeeping, mathematics, and surveying at one of the many academies open in the eastern states during the 1830s. Trained as a civil engineer and as an attorney, he may have apprenticed to an engineer and read law in an attorney's office. Intelligent, restless, and ambitious, Preston personified the successful young male in early nineteenth-century America, "where men start from a humble origin . . . and gradually rise in the world, as the reward of merit and industry."[9] When the Panic of 1837 hit merchants and farmers throughout the East, Preston followed his parents to Illinois.

On August 2, 1838, Preston wed Lucy Ann Hyde at Watertown, New York. The bride brought to the marriage her thirteen-year-old sister Julia and fifteen-year-old brother, George Warren Hyde. Born March 28, 1823, at Batavia, New York, young George Hyde lived with a maternal uncle in Watertown after his father abandoned the family and his mother remarried. The boy attended local schools and

Figure 1 (at left). Pacific Northwest: Oregon and Washington. (Map image by Allan Cartography, Medford, Oregon, 2008.)

Figure 2. John B. Preston, Oregon Surveyor General. (Louis H. Hyde Family Book, Reed-Hyde Papers, Private Collection.)

learned the milling trade in his uncle's flour mills on the Black River.[10] After their marriage, John and Lucy Preston moved to Lockport, Illinois, where he joined the Illinois-Michigan Canal staff as junior assistant engineer. Canal construction begun in 1836 to link Lake Michigan with the Illinois River, pushed forward until 1843 when funding ran out. When work on the project resumed in 1845, Preston, a resident engineer, supervised building the channel's south segment. When the waterway opened in 1848, Preston moved to St. Louis, Missouri, to team with former canal contractor Joel Matteson in the mercantile and forwarding business. Now in his late twenties, George Hyde clerked for a steamboat company running between St. Louis and New Orleans (Figure 3).[11] When Preston received word of his own appointment, he closed out his St. Louis business interests and arranged for his wife, their eight-year-old daughter, and the Hydes to accompany him to Oregon.

The upcoming Oregon surveys beckoned to young men seeking a way west, and during early 1851 hopeful applicants wrote Preston seeking work as his survey assistants. Some correspondents claimed a surveying background. One potential candidate had experience in the surveyor generals' offices of Louisiana and Missouri. In his letter, R. A. Payne confessed to poor health and poignantly expressed the hope "that [Oregon's] mild climate and healthy location would be better adapted to my constitution."[12] Preston cited the intense physical work of surveying and dashed Payne's dream by telling the young

Figure 3. George W. Hyde, Oregon Deputy Surveyor. (Louis H. Hyde Family Book, Reed-Hyde Papers, Private Collection.)

man that his prospects for being appointed a deputy surveyor "were not very bright."[13]

Preston asked Surveyors General Charles Noble of Michigan and Caleb Booth of Wisconsin to recommend two first-class surveyors. The qualifications for selection were specific: they must be experienced surveyors of absolute integrity, ready to leave for Oregon Territory almost immediately, and they must be skilled in operating Burt's Improved Solar Compass. Developed by William A. Burt (1792–1858) after Michigan's iron ore deposits distorted his magnetic needle compass readings, the solar compass used the sun as a fixed reference and employed light instead of the magnetic needle to precisely locate due north. As Burt himself observed, the compass worked "astronomically in determining latitude, and in measuring horizontal angles from the true meridian, and in determining the declination, and hour arcs, of celestial objects within the Zodiac."[14] Philadelphia craftsman William J. Young built Burt's new invention and federal officials patented it on February 26, 1836. In 1850, the General Land Office adopted the solar compass for surveying major boundary lines in regions of magnetic disturbance. Aware that magnetic interference occurred in Oregon, officials required that Preston's surveyors use Burt's elegant, versatile compass for major survey lines (Figure 4).[15]

Summoned by General Land Office Commissioner Justin Butterfield, Preston reached Washington, DC, in February, 1851, with less than a month to collect information about Oregon, purchase and ship technical instruments, and buy stationery supplies for his new office. Preston made three quick trips north in less than three weeks. In New York he bought transits to measure horizontal and vertical angles and sextants to calculate angular distances for latitude and longitude, and in Philadelphia he purchased four solar compasses from William Young.[16] Hurrying back to Washington, the surveyor general met with Dr. John Evans, newly commissioned as US Geologist for Oregon, who would travel overland to the Territory the following summer. Evans asked Preston to carry his delicate barometer tubes, essential for making barometrical measurements, aboard ship.[17]

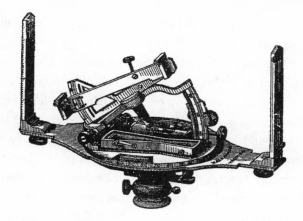

Figure 4. Solar Compass. (Reproduced by permission from Francois D. Uzes, *Illustrated Price Guide to Antique Surveying Instruments and Book.* Rancho Cordova, California: Landmark Enterprises, 1980.)

Preston's time grew tight. Congress appropriated funds for the Oregon surveys on March 3. He met several times during the next few days with Commissioner Butterfield and with Principal Clerk of Surveys, John Moore.[18] Moore handed Preston a sketch map of Oregon west of the Cascades, the ten-year-old plat "compiled from the best authorities accessible to the office at the time it was made." The linen sheet incorporated information gathered by Lieutenant Charles Wilkes, a naval officer and explorer who had visited the Oregon Country in 1841 as part of a federally funded scientific expedition. The map depicted the Columbia River's supposed course and would presumably guide the surveyors in establishing the principal meridian and base lines. Clerk Moore gave Preston several leather-bound copies of his new *Manual of Surveying Instructions*, directed to the "Surveyor General of Public Lands in Oregon." With an exchange of letters between Oregon and Washington taking three to four months, Moore intended the slim volume, the first to consolidate authorized federal surveying procedures into one handbook, to answer questions that would surely arise once the surveys began.[19]

As John Preston dashed between meetings in Washington, two surveyors on opposite sides of Lake Michigan packed for the long voyage to Oregon. Preston chose thirty-three-year-old William Ives of Michigan and James Freeman, thirty-four, of Wisconsin as his lead surveyors. Most likely unknown to each other, Ives and Freeman had solid reputations as accurate, honest engineers. Season after season between 1840 and 1850, the two men had traced the rectangular survey's orderly pattern onto the earth. Across public lands of Wisconsin, Iowa, and Minnesota, they had created a grid "upon which the small freeholds of the ideal agrarian democracy could be laid out like checkers on a board."[20]

When working in any undeveloped country, surveyors first set an "initial point," the intersection of a principal meridian running north and south with a base line running east and west. Using as a guide Davies' standard *Elements of Surveying and Navigation*, a text favored by government surveyors, Ives and Freeman had measured north-south parallel meridians six miles apart, and east-west horizontal lines that also lay six miles distant from each other. The equal squares they laid out were designated as townships; those townships lying along the same meridian were called a range. The surveyors numbered townships and ranges outward from the initial point; the first township located north of the baseline and east of the meridian, for example, becoming Township 1 North, Range 1 East (Figure 5). They further divided each of the checkerboard's townships into thirty-six square-mile sections, numbering them sequentially from the northeast corner to the southeast corner (Figure 6).[21]

William Ives and James Freeman had left their parents' eastern farms in the late 1830s seeking survey work along the expanding frontier where intense agriculture and timber speculation drew settlers to newly opened federal lands. The younger of the two men by a year, William Ives was born April 10, 1817, in Sheffield, Massachusetts. The year of his birth, his father moved the family to the Housatonic Valley where he farmed land between the Taconic Mountains and the Berkshire Hills. In 1839, now twenty-two years old, tall, and described

16

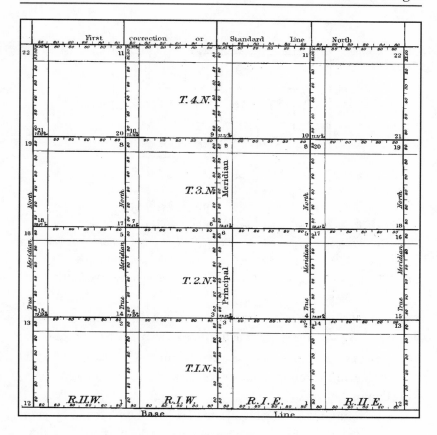

Figure 5. Township order north of the base line; east and west of the meridian.
(Reprinted from John M. Moore, *Instructions to the Surveyors General of the
Public Lands for the Surveying Districts Established In and Since the Year 1850.*
Washington, DC: A. O. P. Nicholson, Public Printer, 1855.)

by one family member as "not handsome but . . . good and kind-
hearted," young Ives left Sheffield, Massachusetts, for Detroit, Michigan.[22] He soon found steady work as a chainman on William Burt's
crew for surveys on Michigan's northeastern Lower Peninsula in the
1840 season, quickly mastering the solar compass. At Burt's recommendation, Michigan's surveyor general named Ives a US deputy surveyor in March, 1843, thereby qualifying the young man to take survey

6	5	4	3	2	1
7	8	9	10	11	12
18	17	16	15	14	13
19	20	21	22	23	24
30	29	28	27	26	25
31	32	33	34	35	36

Figure 6. Township with numbered sections. (Reprinted from *Manual of Surveying Instructions for the Survey Of the Public Lands of the United States and Private Land Claims.* Washington, DC: Government Printing Office, 1902.)

contracts of his own. As Burt's compassman on Michigan's Upper Peninsula in 1844, Ives shared in the discovery of the first of the Great Lakes iron ranges and in 1845 assisted his mentor in measuring 4,000 miles of line in the Portage and Sturgeon lakes areas of northern Michigan (Figure 7).[23]

Ives returned to Detroit from the Huron Mountains in November, 1846, to learn of his parents' simultaneous deaths from infectious disease contracted on the train home to Massachusetts after a visit to Michigan. In that bleak winter, the surveyor faced his loss with sorrow. As the weeks went on, however, he enjoyed some social distraction. He writes, "From the 12th of January until the 10th of February, I did not accomplish much. I copied a few maps, read Geology etc. and was unwell 8 or 10 days partly from the affects of a sleigh ride . . . where the whole company got drunk & had what was called a merry time."[24] In the spring of 1847, Ives signed his first solo contract to survey Isle Royale on Lake Superior, and in subsequent seasons led expeditions to Milla Coquins River, Grand Island, and Lake Superior.[25]

William Ives asked his youngest brother, twenty-one-year-old Butler, to work with him in Oregon. Born in Sheffield, Massachusetts,

Figure 7. William Ives, Oregon Deputy Surveyor. (Courtesy of State Archives of Michigan, Lansing; original image is from the William Ives Gilchrist collection.)

January 31, 1830, Butler Ives moved to Detroit after his parents' deaths and, from 1848 to 1850, apprenticed to his older brother on survey contracts, gaining his own appointment as Michigan deputy surveyor on May 3, 1850. In December, after the survey season ended, Ives enrolled in the Lodi Plains Academy southwest of Ann Arbor to study Latin and Greek in preparation for entering the University of Michigan.[26] When Butler Ives opened his brother's letter with the news about Oregon, he abandoned his immediate educational plans. On March 10, 1851, he took the train to Detroit, arriving just in time to collect $400 from his father's estate and give his brother, Caleb Ives, his power of attorney. Two days later, he left Detroit for Buffalo and a short visit to the old family home in Sheffield, Massachusetts, before meeting William Ives in New York.[27]

⤸

James Eldridge Freeman learned of the Oregon surveys while at work in Wisconsin. Born March 19, 1816, in the rolling hill country of New York's southwestern corner, young Freeman helped his father farm the clay loam soil near Lake Chautauqua.[28] Crossing Lake Erie by steamer in 1840, he struck out for Wisconsin Territory where rising land values drew eager settlers. In subsequent seasons, Freeman surveyed in southeastern Iowa and in the upland country of southwestern Wisconsin between the Black and Trempeleau rivers. Appointed a US deputy surveyor for Wisconsin, Iowa, and Minnesota, Freeman endured freezing winters and dank, insect-infested summers to run his lines.[29] Surveying townships in pine, fir, and birch timbered Isanti and Kanebec counties of eastern Minnesota Territory in the fall of 1849, Freeman observed the already widespread exploitation of Midwest forests. "Much of the pine is already disappeared, and much more will disappear this and every succeeding winter, before the companies of 'log-gers' who [stop] wherever they can find a grove or skirt of this timber of such size as to occupy them during the winter."[30]

At the half-century, with Michigan and Wisconsin largely settled and California gold shifting the nation's economic focus westward, Freeman's contract opportunities thinned. He took a job as Grant County surveyor in southwestern Wisconsin where he boarded with a family in Lancaster.[31] When Surveyor General Caleb Booth recommended him for work in Oregon, the surveyor resigned his county post. Like William Ives, James Freeman sought new professional standing in Oregon and, as a result, a wide choice in survey contracts with little competition from others. Most of all, he counted on reaping some healthy profits.

⤺

On March 13, Preston met several new recruits on the busy New York pier.[32] Allen McH. Seymour, brother-in-law of Thomas Nelson, newly appointed Chief Justice for Oregon Territory and also going aboard ship, appeared with Zenas Ferry Moody and John Stephenson. All three youths had gained crew positions on the Oregon surveys through political connections. Allan P. Millar, who, with his wife, was a close friend of the Preston family, also joined the group. Preston had recently appointed him as an assistant clerk in the surveyor general's office. James Freeman also joined John Preston on the dock. At thirty-four, Freeman stood apart as the one man in Preston's entourage to win his place through expert qualifications rather than social connections.[33] Despite their differing social and professional credentials, however, the outgoing Freeman would forge enduring relationships with Preston and his more privileged companions during the voyage.

The Prestons and other Oregon-bound passengers gathered for roll call on the side-wheel steamer *Empire City*, one of the most luxurious steamers on the water. The vessel's upper deck housed staterooms, an elaborately furnished women's reception room, and a smoking lounge for men, while additional staterooms flanked the dining saloon on the main deck.[34] As the steamer headed out to sea, Preston and Nelson joined attorney Stephen F. Chadwick and Samuel R. Thurston, territorial delegate to the US Congress, in the lounge. Returning home to Oregon at the end of his term in office, Thurston chaperoned five young women aboard ship. Sponsored by the Congregational Church Society of Vermont, they had been hired by Oregon City minister George H. Atkinson to teach at the Clackamas County Female Academy in Oregon City and the Tualatin Academy west of Portland.[35]

On March 22, the passengers disembarked the *Empire City* at Chagres on the Caribbean shore of the Isthmus of Panama. This part of the trip over, the travelers crossed the sixty-mile-wide Isthmus on barges towed upriver by steamboat. There, the travelers had to wait ten long days in a Panama City hotel until passengers aboard another steamer from New York could join them for the next leg of the journey to San Francisco.[36] Preston wrote General Land Office Commissioner Justin Butterfield in Washington that everyone was well except for attacks of seasickness and that most of the equipment had arrived safely except for one of Dr. Evans' barometers, broken by a falling mule.[37]

The passengers departed Panama for San Francisco on April 2 aboard the steamer *California*. As the overcrowded vessel neared Acapulco a week later, thirty-eight-year-old Samuel Thurston suddenly died. His fellow passengers were unaware of the severity of the Oregon delegate's illness because, as Thomas Nelson remarked somewhat uncharitably, he had "exhibited so much impatience, petulance, and . . . selfishness that no one believed that he was really sick until the day before his death." Thurston's body lay on deck overnight wrapped in an American flag. When the steamer reached Acapulco the next day, the Prestons and other passengers walked behind the casket as hired men carried it to the burial place in rocky soil.[38] The realization that Thurston's widow and the Territory could not know of his death for several weeks subdued the Oregon-bound passengers. John Preston himself fell ill with a fever, but quickly recovered. He and Lucy Preston now looked after the young women who had lost their chaperone.

The *California* reached San Francisco on April 23, and the Prestons took rooms at the Union Hotel, enjoying parlors with plush carpets, elegant furniture, and sumptuous meals.[39] While in the city, Preston called on Robert Elder, who for twelve years had been an assistant engineer on the Illinois-Michigan Canal before coming to California in 1849. Born in Scotland about 1812, Elder immigrated to the United States during the 1830s. Preston offered Elder a clerk's position in his office and the engineer agreed to come to Oregon soon.

The Prestons and their entourage left San Francisco the next day aboard the Pacific Mail Steamship Company's new vessel *Columbia*. Built especially for the Oregon route, the 777-ton side-wheeler made the four-day voyage from San Francisco north twice a month. The ship reached Astoria, Oregon, at the mouth of the Columbia River, four days later on April 28.[40] The passengers bound for Oregon City boarded the small vessel *Lot Whitcomb* and steamed up the Columbia River to dock for the night at the US Army post at Fort Vancouver. An important fur depot, trading center, and political headquarters since its founding by the Hudson's Bay Company on the Columbia's north

bank almost thirty years earlier, the compound remained a key location in the Oregon Country. Early the next morning, the passengers
gazed out over a dozen buildings lining the parade ground, all enclosed by a tall picket fence and set against a striking background of
"mountain peaks . . . rolling hills, winding river and green lawn."[41]

The *Lot Whitcomb* navigated down the Columbia from Vancouver
and entered the Willamette River where, at the next stop, the Preston
family and the teachers briefly disembarked, the young women walking "up from the steamer's plank through a double line of gazers composed of the entire population of Portland."[42] Proceeding upriver as
far as it could go, the steamer stopped at Milwaukie where the two
dozen Oregon City passengers boarded a flat, open "whale boat" for
the rest of the trip upriver (Figure 8).

Late in the afternoon, in the rapids at the confluence of the
Clackamas and Willamette rivers, the craft lodged fast on a submerged
bar. When the oarsmen's efforts to free the boat failed, the male passengers waded ashore and lit a huge bonfire. The women remained
aboard the boat and huddled under blankets in the open air.[43] Later
that evening, Nelson and two other men left their companions to watch
over the stranded passengers and struck out for Oregon City. Hiking
several hours over a rough trail through woods and brush and wading
across sloughs, they reached town at midnight. After first telling the
news of Samuel Thurston's death, the worn-out travelers collapsed in
sleep at a hotel.[44] The next morning, a relief party rowed downriver to
rescue the stranded passengers. They shuttled them across the
Clackamas River by canoe, and from there the bedraggled travelers
walked the rest of the way to Oregon City. When they arrived, the
Prestons delivered the teachers into the Reverend Atkinson's care.[45]
The surveyor general settled his exhausted wife and daughter at the
Main Street House, a two-story hotel offering the town's best accommodations, and walked out to explore the town.

Established at the falls of the Willamette in 1842 by Hudson's
Bay Company's former Chief Factor John McLoughlin, Oregon City
lay twelve miles upriver from Portland on the east bank, opposite

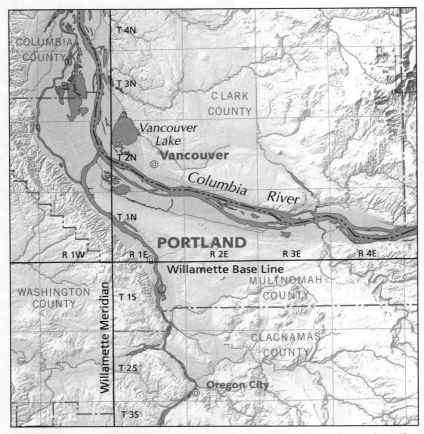

Figure 8. Portland and Oregon City, Oregon, vicinity. (Map image by Allan Cartography, Medford, Oregon, 2008.)

Linn City (now West Linn). Named the territorial capital in 1848, the town stretched between a steep bluff and the river (Figure 9). One visitor saw it as a "bustling little village [with] neatly painted white houses," while another called it "a miserable cramped-up place situated in a rocky canyon."[46] James Freeman compared the town and the nearby settlements to ancient Rome, observing "that there [stood] one city on seven hills, while here are seven cities on one hill."[47]

In Oregon City's oldest section, Preston walked past wood frame commercial buildings and the Congregational and Catholic churches.

Figure 9. Oregon City, Oregon, 1857. (Lorain Lorenzo Photograph, Oregon Historical Society, OrHi 21079.)

He visited the *Oregon Spectator* newspaper office at Fourth and Main and the territorial statehouse at Main and Sixth Streets. Short cross-streets that reached from the bluff down to the river intersected Center, Main and Water streets, while atop the bluff newer dwellings flanked roads laid out by surveyor Jesse Applegate in 1848.[48] Accustomed to the stone and brick commercial buildings of Chicago and St. Louis, John Preston thought Oregon City's small wood frame buildings and muddy streets were very rough indeed.

Within a few days of his arrival, the surveyor general called on Governor John P. Gaines.[49] A Whig, Gaines had succeeded Democrat Joseph Lane, appointed by President James Polk in 1848 as the territory's first governor. Inaugurated in Oregon City on March 3, 1849, Lane had served briefly, resigning upon Zachary Taylor's election as President. Taylor then appointed Gaines, who took office on August 18, 1850. By the time John Preston arrived, Gaines already faced vocal critics. "Everything was against him," one historian noted. "He was a Whig, while Oregonians were Democrats; he was dignified — pompous according to his enemies — and a marked contrast to the easygoing frontiersman, Joe Lane. In less than a year Gaines was at swords' points with the majority of Oregonians."[50]

Shocked at Oregon City's inflated prices, Surveyor General Preston wrote Commissioner Butterfield that the departure of most of the town's male residents to the southern Oregon mines had caused "wages and the necessaries of life to be exorbitantly high." In addition, Preston predicted, the town's decimated population was not likely to yield many competent men to assist the deputy surveyors. The surveyor general accounted for costs to date: $2500 for his own annual salary; $2200 for each of two clerks' annual salaries; $1800 for a third clerk, and additional pay for a servant and a messenger. Finally, Preston summarized his own travel expenditures from St. Louis to Washington to New York and Philadelphia, and to Oregon, plus the cost of equipment and supplies at $4,680.74.[51]

Preston sought appropriate quarters for his office, heeding federal regulations that the building stand well apart from other structures to minimize risk of fire. John McLoughlin offered him use of an empty dwelling on the east side of Main Street for fifty dollars a month.[52] McLoughlin, born in 1784 in Quebec, Canada, had served as chief agent and administrator of the Hudson's Bay Company in the Pacific Northwest from 1824 to 1846. Headquartered at Fort Vancouver after 1825, McLoughlin helped French Canadians settle the Willamette Valley and later sheltered and fed American missionaries and settlers who ventured into the region. After leaving HBC's employ, he moved to his Oregon City claim early in 1846 and rapidly acquired valuable residential, commercial, and industrial holdings in town.

Preston paid a deposit on the rent, bought a desk, tables, and stools, and set Hyde and Millar to organizing the office. The clerks arranged the drafting tables in the parlor and created a separate office for Preston in an adjoining room. They stored the field books, stationery, and mapping paper in cupboards and stacked the firewood behind the building. Settled at his desk, Preston tackled the pile of mail that had accumulated at the post office prior to his arrival. Settlers south of the Columbia River had questions about their land claims. At

Olympia, near Puget Sound at the end of the overland route from the Columbia and a focus for settlement, other emigrants wanted to know when the public land surveys would begin.[53]

As he waited for William Ives to reach Oregon City, the surveyor general reviewed his instructions from Washington. The General Land Office advised him to establish the initial point — the place at which the Willamette Meridian and Base Line intersected — so that the base line running east would cross south of the Columbia River and north of Mount Hood and running west would cross the Tualatin Plains. The Willamette Meridian, the Commissioner directed, should pass between Vancouver Lake and the Columbia River to keep the line out of these two waterways. The meridian's northerly terminus was set at Puget Sound.[54]

Although the General Land Office planned eventually to extend the Willamette Meridian to the Oregon-California boundary, for now Butterfield instructed Preston to survey only the Willamette and Umpqua valleys' rich farmlands, discounting southwestern Oregon as "too unattractive to make the surveying of it available at present." After conferring with McLoughlin and other seasoned Oregonians, Preston wrote Butterfield, "Owing to the uncertain information about the position of the Columbia river, it will be necessary for me to go to the Cascade mountains, and have a random [trial] line run from the most southerly point in the River west to the meridian line (or random from the mouth of the Willamette) . . . to designate the starting point for the Base and meridian lines."[55]

∽

Still in Detroit as the Prestons left New York, William Ives enlisted twenty-six-year-old Joseph Hunt and thirty-year-old Robert N. Brevoort to work for him in Oregon. Robert Hunt, a relative of Joseph Hunt, also joined the group.[56] Ives carried letters of recommendation to give to Preston in Oregon. Charles Noble called him "an intelligent, active, laborious & faithful surveyor." Geologist Bela Hubbard thought him "one of the most competent & faithful surveyors ever employed in that region." Describing Ives as a man of character and integrity, William Burt recommended him enthusiastically as a "competent, efficient and faithful surveyor."[57]

Ives left Detroit by steamer for Buffalo on March 17, 1851. In New York City five days later, he joined his younger brother, who had just arrived. The brothers paid $50 each to share a cabin on Howland & Aspinwall's *Cherokee* and left New York on March 28 amidst a mass of spectators gathered to see the steamer off. Just before the men boarded ship, friends burst out of the cheering crowd with four bottles of brandy and two bottles of wine. The excitement of departure, however, soon gave way to gut-wrenching seasickness. On April 3, partly recovered from five days of misery, the brothers watched from the deck as their ship passed the *Empire City* on its return run to New York from Chagres.[58] The *Cherokee's* three hundred passengers settled in for long, tedious days, during which, Butler Ives observed, they had nothing to do "but eat (which was sometimes hard on account of seasickness) . . . sleep & lounge about the boat which was mighty hot."[59]

The *Cherokee* reached Chagres on April 7 and two days later steamed up the Chagres River to Gorgona. Joined by his former surveying assistant Loren L. Williams, William Ives started out on foot for Panama with Joseph Hunt, leaving the younger men, Butler Ives and Robert Hunt, to lug the baggage.[60] The party left Panama April 15 aboard the *New Orleans*. William Ives spent $225 for a tiny, crowded cabin with a porthole, while Butler Ives and the other men paid $100 each for berths deep in the dark, noisy, steerage. Two days out of Acapulco on April 20, *New Orleans'* employees hauled a crewman's corpse onto the deck. Unlike the ceremony in which Samuel Thurston's body lay enshrined for a night in the "stars and stripes," no honors attended this man's burial. His colleagues sewed the body into a canvas shroud and strapped a bag of coal to the ankles. After the captain read a short service, the men lifted the heavy bundle onto a plank hanging over the side of the ship and tipped it into the sea. "But little feeling seemed to be manifested by the circumstances," Butler Ives wrote of the bleak event, "and within one hour they all appeared to be following the general run of conversation as usual."[61]

The *New Orleans* reached Acapulco on April 22 at midnight. The next morning Ives and his companions faced a medical crisis of their own. Loren Williams fell gravely ill, and the ship's doctor announced that Williams could not possibly survive aboard the vessel. The ailing man's friends then carried him ashore and left him in the care of an American doctor. Robert Hunt volunteered to stay with Williams until he recovered and they could board another ship, and William Ives gave Hunt money to cover the sick man's expenses.[62] The next day, Ives and his men left Acapulco for San Francisco.

⤶

On May 6, in the wake of the spectacular fire that heralded their arrival in San Francisco, William and Butler Ives, Joseph Hunt, and Robert Brevoort left the still-smoldering city aboard the *Columbia* with a few other passengers and the US mail.[63] Four days later the steamer crossed the Columbia River bar in heavy surf to dock at Astoria twelve miles above the river's mouth. Later in the evening as the vessel moved upstream in deepening darkness, the surveyors strained to make out the landscape. Butler Ives observed, "I was unable to see much of the scenery along the river but what we did see was very rough & appeared thickly timbered & uninviting."[64]

As the steamer approached Portland at dawn on May 11, the brothers awakened as the ship's thundering cannon signaled their arrival and the end of hard weeks at sea. Lying between the Willamette River and dark, forested hills beyond, the town was crowded with new, hastily built frame buildings. After a simple breakfast at the Columbian Hotel, the men hired a boatman to take them partway upriver. Aching for exercise, William and Butler Ives got out of the boat at the town of Willamette and, as Preston and Freeman had done ten days earlier, entered Oregon City on foot.[65] The brothers retrieved their trunks at the dock and hauled them up to the Main Street House before heading for the surveyor general's office. After introducing themselves to

Preston, the two set out to explore Oregon City. "Mr. Preston & his clerks say the villagers mostly had gone to the gold mines & left the place rather desolate," Butler Ives observed, adding that the country itself was bleak, "hilly & rocky & not very inviting."[66]

The brothers tried to reconcile this raw landscape with the one they had imagined. So unlike the familiar terrain of Michigan and their native Massachusetts, these jagged, high mountain peaks and dark blankets of towering evergreens astonished the newcomers. Just as overwhelming as the vistas were the uncertainties that lay ahead: William and Butler Ives would have to prove themselves to the new surveyor general and outfit survey expeditions in a country where they knew no one but each other. Two months and thousands of miles from home, they faced trials as severe as any in the past — ordeals that would shape their lives for years to come.

Chapter 2 – Summer, 1851

On Line in Oregon Territory

" We are all well, minus parts of our pantaloons."

Preston gave William Ives little time to rest, instructing him to report for work in less than thirty-six hours. Keen on burnishing his own reputation, the surveyor general was eager to launch his expedition to examine the Columbia's route upstream. On May 13, US deputy surveyors Ives and Freeman, strangers and competitors each a bit wary of the other, joined Preston aboard a steamer bound for Fort Vancouver to begin the survey trip upriver. Accompanied by Butler Ives and Joseph Hunt, the party reached their destination in midafternoon. Once again, the strikingly sited Fort Vancouver impressed Butler Ives: "The Hudson's Bay Company have large & beautiful claims here besides a fort or enclosure where they have a building established for trading with the Indians." Seeing Vancouver at close range, he judged it "one of the most pleasant & handsome places yet noticed in Oregon" (Figure 10).[1]

Early the next day, Preston, William Ives, and Freeman hired a Hudson's Bay Company batteau for a three-day trip up the Columbia. On this mild May morning, the men stared out over the broad plain. On each side of the river lay country staggeringly different from any terrain they had known before. Dark, timbered ridges etched the horizon

Figure 10. Hudson's Bay Fort, Fort Vancouver, Columbia River. (Watercolor by Mr. Kashnor, [1927–1928], based on sketches by Trevenen Penrose Coode, [1843–1847], HBCA P-133; Hudson's Bay Company Archives, Archives of Manitoba.)

and ahead in the distance loomed the steep slopes of the Cascades. Accustomed to mountains of modest elevation, the engineers gaped at soaring, snow-covered Mount Hood. As the boatman powered the craft upstream, his passengers studied the Columbia's course, noting the location of its southernmost bend north of present-day Corbett. Going ashore to calculate their position, the surveyors ensured that the base line would run well south of the Columbia channel as well as pass conveniently near the mouth of the Willamette River.

On May 17, the oarsman turned the boat for the fifteen-mile return trip downstream. Back at Fort Vancouver, the explorers found Butler Ives and Joseph Hunt trying to adjust the solar compasses in cloudy weather, a tedious, frustrating task that would perpetually plague the surveyors in Oregon.[2] James Freeman, who had developed a collegial rapport with Preston, left for Oregon City with him the following day. William Ives bought a skiff for the next leg of the survey expedition, and on May 20, with his younger brother and Joseph Hunt, he left Fort Vancouver in the skiff to establish an optimum location for the meridian line. They boated down the Columbia to the Willamette River's mouth and, jotting measurements there, hiked a mile-and-a-half north. Then, to be sure that the meridian would pass along the west side of a large lake (Vancouver Lake), the men surveyed a temporary line back south to the Columbia. Sitting in front of a settler's warm fire that night, William Ives wrote his daily journal entry: "Ran a random [temporary] line for the first principal meridian in Oregon Territory south of the Columbia River."[3]

Over the next two days, William and Butler Ives extended the line south across the Columbia, portaging around small lakes and swamps, and reached the Willamette. They crossed the Willamette River and hacked through the heavy brush to survey the meridian line south one more mile to a hill. Finishing this initial reconnaissance, the men hiked back to Oregon City. Practicing the thrift that would be the hallmark of all their good-weather, between-contract waiting periods, they camped on the west bank of the Willamette.[4] Well-equipped with tents and cooking utensils, the Ives brothers camped at various locations, but usually favored the Linn City side of the river. Here they competed for sites with the laborers, miners, and other men who all slept out to save money. Theft was occasionally a problem. The day after William Ives returned from the meridian survey, Butler Ives and Joseph Hunt took his new skiff across the river to attend church. When they returned, the boat was gone. A frantic search of the area turned up nothing. The young men hunted two days for the craft — William Ives had owned it only for five days. They hiked downstream as far as Milwaukie and took a steamer to Portland to search along the bank for the boat, but "found it not."[5]

On May 28, 1851, as his two deputies stood before him, Surveyor General Preston gave James Freeman Contract 1, the first federal survey contract in Oregon, "to faithfully survey the Willamette Meridian from its intersection with the Base line south to the Umpquah Valley," an estimated 210 miles. He awarded William Ives Contract 2, projected at 240 miles, "to run the Base line from its intersection with the Willamette Meridian, west to the Pacific Ocean and East to the summit of the Cascade Range of Mts and also to run the Willamette meridian north of the Base Line to Puget Sound" (Figure 1).[6]

Preston handed each man a copy of the map given him in Washington that embraced "all the information in the possession of this office." Preston directed the deputies to return the field notes of their finished contracts by September 25, 1851. General Land Office regulations stipulated that contracts should not encompass more area than could be surveyed within three months, or if the survey district were a long distance from the surveyor general's office, four months.[7] Using government rates for meridian lines, the contracts offered Freeman and Ives $20.00 per mile "for every mile and part of a mile actually run and marked in the field, random lines and offsets not included."[8] The deputies swore to faithfully execute their contracts and posted their bonds. About double the contract amount, a bond ensured a surveyor's honesty and covered the costs should he default.[9] Ives and Freeman financially backed all their expeditions and only after Preston certified that the contract had been satisfactorily completed, could they be paid.

The deputies reviewed General Land Office Clerk Moore's manual for variations from the standard surveying instructions they had used in Michigan and Wisconsin during the previous decade.[10] The Oregon manual appealed to the surveyors' integrity. "The *conscientious* DEPUTY, and his *faithful* ASSISTANTS, however remote and secluded from the eye of human surveillance, will, nevertheless, ever keep in view of the USES, extending into the indefinite *future*, which *their* daily toil is designed . . . to accomplish for the *public good*."[11]

Freeman and Ives must, the manual declared, neatly keep a detailed record of everything noted in measuring their lines. The volume admonished, in one sweeping charge, that they also present "a full and complete *topographical description* of the country surveyed, as to every matter of useful information, or likely to gratify public curiosity."[12] Moore instructed Ives and Freeman to follow special instructions attached to their contracts. Drafted in painstaking detail, the special instructions addressed problems such as wide rivers or inaccessible mountain ranges that made extending the line difficult.[13] Furthermore, the surveyors must keep separate field books for the meridian and base lines; for standard parallels, which are lines introduced at regular intervals to counteract errors resulting from converging meridians; and for the township and section lines.[14]

To gain information about the Oregon Country, the Donation Land Act required use of the geodetic survey as well as the rectangular survey. For mapping the geodetic survey the deputy used the plane table, a specialized device of surveying and drafting equipment, to document distinctive topographical features while he extended lines between widely scattered settlements.[15] Although General Land Office officials subsequently rejected the geodetic system as unwieldy and expensive, Preston, who wanted an accurate description of the region's topography, instructed Freeman and Ives to keep both geodetic and linear field books.[16] He further admonished his surveyors to make precise measurement of the meridian line, "knowing that on the truthfulness will depend the value of the Surveys."[17] At the heart of the surveyor general's directive lay the basic survey issue of magnetic declination or variation — the difference between the true north and magnetic north poles, or "the angle which the magnetic meridian makes with the true meridian, at any place on the surface of the earth."[18] Because magnetic declination varies by location on the Earth's surface, and even at the same place also varies over time, the surveyors had to take variation into constant account for accuracy in surveying all their lines.

William Ives and Joseph Hunt left Oregon City for Portland on May 29, while Butler Ives followed on the *Lot Whitcomb* with their baggage. On May 31, the first sunny day in over a week, the men returned to the ridge south of the Willamette River to survey the temporary meridian two miles farther south.[19] The men scrambled through ravines choked with fallen timber, pushing toward the intended initial point. On June 1, they scaled a hill in view of the Willamette River and "set [a] post intended for the intersection of Base Line & Willamette Meridian, of the Surveys of Oregon." Butler Ives surveyed a preliminary location for the base line three and one-half miles east to Portland, while Hunt went back to set up tents and cook supper. When the surveyors returned for the night they found several of James Freeman's men visiting their camp.[20]

William Ives assembled his crew on June 3, hiring James Swinhart and Charles Smith as axeman and flagman, and signing William A. McDermott, John H. Jones, J. M. Clark, and Nicholas Coarser as chainmen. Nicholas Coarser and James Swinhart could read and write. Charles Smith was illiterate.[21] Ives named his brother as chief axeman and Robert Brevoort as head chainman; as leaders they would train the less skilled employees. Young and physically strong, most of William Ives' employees had come west to find gold. Eventually disappointed in the search, many of the disillusioned young miners looked for work in Oregon. Depending on their duties, Ives paid his men between $60 and $80 per month. Although Willamette Valley farmers offered more money — $5 to $6 a day for harvesting grain — the deputy had no trouble hiring men who preferred surveying to drudgery in the wheat fields. [22]

In Portland, Butler Ives and the rest of William's men stood before a justice of the peace to pledge their honesty and responsibility. The chainmen swore to "level the chain in measuring over uneven ground, and well and truly plumb the tally-pins whether sticking or dropping the same." Butler Ives and his fellow axemen vowed to "mark correctly and distinctly the letters and numbers of all corners." Required at the contract's beginning, at the completion of townships, and at the contract's conclusion, these oaths underscored the crucial need for integrity on the surveys.[23] Nine men, most unknown to each other, circled the fire in camp that night. By the time they emptied the tinned plates heaped with roasted beef and rolled their wool blankets onto the damp ground, they called themselves a company.

Willamette Meridian North, June 4 to June 11, Eighteen Miles

William Ives awoke at dawn June 4 to steady rain. As the crew collapsed the tents and packed bedding, Surveyor General Preston arrived in camp. After breakfast, the packers led the mules north to set up their evening camp, while Preston, William Ives, and James Freeman hiked across the hillside to the initial point. Encircled by onlookers, the deputies pulled out the temporary marker and pounded a stout cedar stake at the intersection of the Willamette Meridian and Base Line. With several blows from the mallet, men secured the first corner for the linear surveys in Oregon.[24]

Ives and Freeman marked the initial point for the fifteenth principal meridian established at the time, adding to what would become a network of meridians and base lines associated with the United States public land survey. Eventually, surveyors would designate thirty-seven initial points for numbered or locally named meridians and base lines in states subject to General Land Office surveys in the continental United States and Alaska.[25] The wild hillside on which the deputies drove Oregon's stake now lies in Portland, enclosed in the small Willamette Stone State Park off NW Skyline Boulevard at a point about three and one-half miles west of the Burnside Bridge.[26]

William Ives headed north from the initial point on the meridian survey. To make sure that every mile was measured accurately, he used two sets of measuring chain, "the foremost [men] using the four pole chain of sixty-six feet, the latter using the two pole chain of thirty-three feet, and the two comparing their measurements at every half mile, and mile station."[27] As the company moved into position, the flagman led the way along the survey line guided by Ives' directions. Through timber and open fields, the flagman searched the terrain for clear paths, marked the line with cloth strips, and doubled back to guide the rest of the party. William Ives operated the solar compass, recorded the calculations, and noted locations of all survey markers and topographical features.[28] Axemen Butler Ives, Swinhart, and Smith followed the deputy surveyor. They stripped away bark to blaze the bearing trees, notched section trees, hewed posts, and inscribed them to mark critical points along the line.[29] Brevoort, McDermott, and the other chainmen came next, unfolding the heavy wire sections of bundled Gunter's chain's and checking them for clogged debris. Then, stretching the chain taut for accuracy, they measured the line.

Invented by Edmund Gunter in the early 17th century, the four-pole chain had 100 links (7.92 inches each) for a full length of sixty-six feet. The two-pole chain had 50 links for thirty-three feet. "The length of every line you run," the Oregon manual directed, "is to be ascertained by precise horizontal measurement, as nearly approximating to an air line as is possible in practice on the earth's surface." To achieve this goal, the chainmen kept the chain horizontally level and stretched as taut as possible over even and rough ground. When handling the chain on steep slopes, the surveyors reduced the chain to half its length in order to retain its horizontal, taut position. Finally, the chainmen had to make sure the tally-pins were plumbed (dropped) accurately to precisely mark the spot they should be stuck.[30] In using the four-pole chain, the rear chainman sighted from his staff to the flag on the foreman's staff, and for the two-pole chain, the flag on the near chainman's staff would serve as a sight object.[31]

The chainmen carried a set of eleven steel chaining pins not longer than fourteen inches long, weighted at the tip and with a ring at the top.[32] Using the tally system, the crew began measuring the line by marking the starting point with one pin in the ground. Front chainman Brevoort wore ten pins suspended from his belt. As he pulled the chain out, the others straightened it and held it up level. Rear chainman McDermott held the rear link mark over his pin and called out, "Stick!" In the lead, Brevoort dropped a pin from zero, stuck it in the ground and shouted, "Stuck!" Advancing by inchworm increments, McDermott and Brevoort repeated this process ten times to measure five chains or one tally of the 50-link chain.[33] The flagman marked the ending positions of the Gunter's chain each time the chainman set its position.

Brevoort wore a tally strap on his belt. A leather cord with eight knots passed through a hole in the strap. When the team measured one tally, the chainman wearing the tally strap pulled a knot through the hole to record "tally one." Rear chainman McDermott switched places with Brevoort at the front and they repeated the steps until they had pulled all eight knots through the tally strap at the close of one-half mile.[34] The chainmen alternately changed places, "each setting the chains he has taken up, so that one is forward in all the odd, and the other in all the even tallies," a procedure intended to prevent mistakes in measurements.[35] Depending on how the deputy assigned tasks, either chainmen or axemen set the posts for the township, section, and quarter-section corners.[36]

The party chopped through brush in the Tualatin Mountains (now known as the West Hills) to run the line to the Willamette River (Figure 11). "All of the land on the last six miles except the bottoms," Ives observed, "has been burnt over and about one third of the timber killed." Burned by lightning or by intentional or accidental human-set fires, areas of scorched trees choked with fallen limbs and debris spread across much of the wooded terrain through which the surveyors passed. By late afternoon, the company neared the river by crossing through Milton Doane's fenced, cultivated field. Settler Doane's log house, "filled with his wife and children," stood nearby. The party crossed the Willamette and camped on the north bank where Preston and Hunt joined them for the night. The surveyor general accompanied the survey the next day over swampy wetlands to the Columbia River, eight miles north of the initial point. That night, the company camped among cottonwoods and willows on the westerly end of what is now Hayden Island.[37]

Early on the morning of June 6, Ives and Preston refigured the distance over the Columbia using a theodolite, similar to a transit but more precise in measurement (Figure 12).[38] Preston, Ives, and Hunt calculated a triangle with the instrument, establishing two points on the river bank and one point on the island. Measuring the distance between the two points on the bank and figuring the horizontal angles, the surveyors gauged the distance over the river to the island. They then measured across the island, figuring the distance from it to the

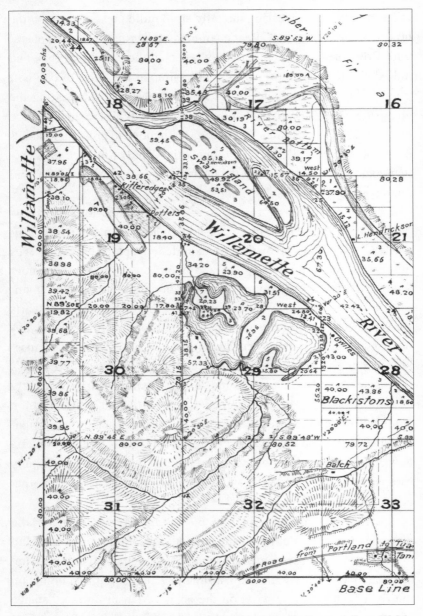

Figure 11. General Land Office Map: Township 1 North, Range 1 East, W. M.; Willamette Meridian from Initial Point to the Willamette River, Contract 2, William Ives, 1851. (USDI Bureau of Land Management, Portland, Oregon.)

Figure 12. Theodolite. (Reprinted from Charles Davies, *Elements of Surveying and Navigation, with Description of the Instruments and the Necessary Tables.* New York, New York: A. S. Barnes & Co., 1854.)

Columbia's north bank. From this point, Ives ran the meridian two and one-half more miles north to see it pass west of Vancouver Lake (Figure 13). His task completed, Surveyor General Preston returned to Portland where he stopped at the *Oregonian* office to announce the beginning of the Oregon public land surveys. Enthusiastic that the work was under way, the newspaper editor cautioned readers, "remember . . . this is a Herculean task, requiring time for its accomplishment."[39]

41

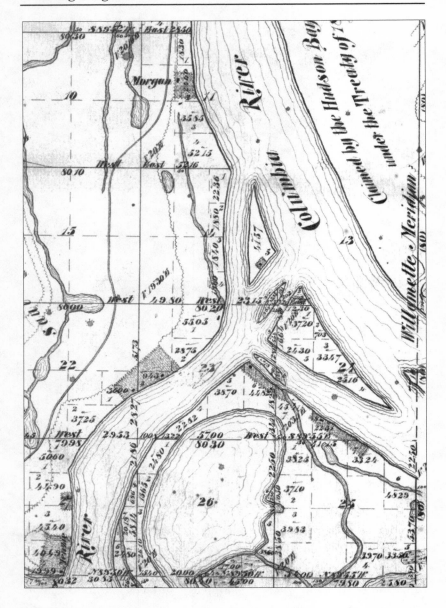

Figure 13. General Land Office Map: Township 2 North, Range 1 West, W. M.; Willamette Meridian over the Columbia River, Contract 2, William Ives, 1851. (USDI Bureau of Land Management, Portland, Oregon.)

William Ives and his men advanced three miles on the east side of the Columbia River on June 7 to reach Shillapoo Lake — the corner of Townships 2 and 3 North lay in the lake — where they camped in a grove of black cottonwoods. The crew slogged around Vancouver Slough on Lake River and East of Campbell Lake, about two miles south of present-day Ridgefield, Washington. Then, two days later, William Ives set the corner of Townships 3 and 4 North, Ranges 1 East and 1 West, eighteen miles north of the initial point.[40] In this swampy country, Ives lacked bearing trees large enough to establish the locations of the meander line posts but assured the surveyor general that he had marked enough points to find the corners later. Measured at the line of mean high water for rivers and lakes, the meander line establishes the course of the banks on a stream or lake. Ives did admit that many of the corners stood "in the water where mounds could not be made nor any thing else done better than what is done."[41]

Frustrated by dense clouds that prevented his using the solar compass, Ives sent his brother, Brevoort, and Smith back to Portland to begin calculations on the base line west of the initial point, while he and the remainder of the company stayed on the line, in case the clouds should clear.[42] On June 10, Butler Ives found James Freeman and his men still camped near the initial point waiting for the sun. The following day, Ives and Charles Smith surveyed a temporary line west along the road leading out to the Tualatin Plains.[43] After determining that the base line would run conveniently through the settled region near Walter Pomeroy's claim, they returned east as far as Freeman's camp.[44]

Base Line East Four Miles, then West Thirty-Seven Miles, June 12 to June 30

William Ives gave up. "Cloudy weather came in," he wrote the surveyor general, "& I took the liberty of leaving." On June 12, he joined his brother at the initial point. Before surveying the base line west, Ives decided first to extend it east to Portland and across the Willamette River. He measured four miles to the western Portland suburbs, drove a quarter-section post inside a chicken house and pushed on toward the Willamette River. Sixty-five years later, a Portland journalist described the base line's location in everyday terms: "The traffic policeman at Broadway and Morrison Streets stands on the line. It passes through the center of the Meier and Frank company building" (Figure 14).[45]

In the late afternoon, Ives crossed the river and camped on the wooded east bank, the site now buried under concrete and interstate bridges. Although the surveyors could not have envisaged this concrete maze, they did anticipate the disappearance of forest land as they observed settlers clearing their fields of trees and brush. Ives set a temporary post east of the Willamette the next day, writing Preston that he would now survey the base line west over the Tualatin Plains, "I shall probably be running through there in three or four days if the weather is as good as usual & perhaps sooner if the sun should shine 16 hours every day. We are all well," he said, "but are minus parts of our pantaloons."[46]

Butler Ives loaded the three mules with tents, blankets, and cooking utensils. He packed the solar compass, magnetic compass, and transit into sturdy boxes and slipped the chain, tally-pins, and field books into separate pouches (Figure 15). Smith and Clark strapped on sacks of beans, dried fruit, rice, and coffee; the survey party would buy fresh beef, pork, eggs, and vegetables from farmers along the line. Heading west from the initial point, William Ives crossed Cedar Mill, Willow, and Rock creeks, north of present-day Beaverton, Oregon, on the base line west.[47] He surveyed the line past David Williams' house and across Michael Moore's field through what are now the residential enclaves of Cedar Hills, Marlene Village, and the sprawling suburbs of present-day Hillsboro. Passing through a stand of timber, the company entered the tiny community of a half-dozen buildings. They chained the base line past the small, log "Hillsborough" courthouse, briefly dispersing the crowd that had gathered to watch a murder trial (Figure 16).

Near that log courthouse in the center of town, William Ives erected the first survey mound in Oregon at the corner of Townships

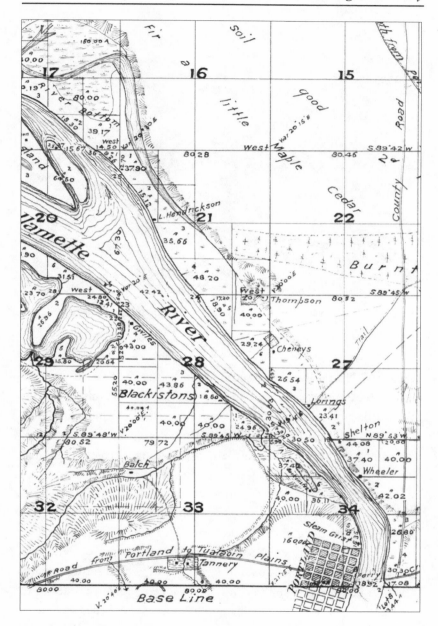

Figure 14. General Land Office Map: Township 1 North, Range 1 East, W. M.; Willamette Base Line through Portland, Contract 2, William Ives, 1851. (USDI Bureau of Land Management, Portland, Oregon.)

Figure 15. Butler Ives, Oregon Deputy Surveyor *(at right)*, and crew. (Private Collection.)

1 South, 1 North, Ranges 2 and 3 West.[48] When bearing trees were unavailable as references for corner posts, the surveyors raised mounds of earth or stone around the posts to insure their permanency. Surveyor General Preston, who had come from Portland for the occasion, watched as Ives shaped the earthen mound to the correct height and encased it in rock — most likely ignoring the manual's somewhat impractical suggestion that deputies mark township corners by planting seeds of a fruit tree "so that . . . a small clump of trees may possibly hereafter note the place of the corner" (Figure 17).[49]

"The Plains," Butler Ives said, were "fine." The party made good time chaining across the Tualatin River Valley where settlers, many of whom had arrived before 1850, had planted fields to wheat and other grains. Cattle, oxen, sheep, and hogs grazed in broad fields enclosed with split-rail fencing near the present-day towns of Cornelius and Forest Grove. As William Ives surveyed the base line across Range 3 West on June 18, Butler Ives and two of the packers led the mules ahead along the "Hillsborough-West Prairie" road, now State Highway 8, its course nearly parallel to the base line through the township. He and McDermott set up camp at "West Prairie" known today as Forest Grove. After surveying the line within a few miles of "Tualitin Plains Academy," today's Pacific University, William Ives joined them for the night.[50]

Deputy surveyor Ives and his men entered the mountains of the Coast Range, crossing Gales Creek several times as they climbed the forested ridge. Butler Ives and two assistants drove the loaded mules ahead to establish camp near Allen's mill, which lay north of the base line and along the creek. On June 20, Ives and Coarser left camp with four days' provisions loaded on one of the mules and struck out after the survey party. Pushing hard, they tracked them four miles into the mountains where deep ravines and thick brush hindered their progress. When the two were unable to overtake the company, William Ives and his men spent the night without blankets or supper.[51]

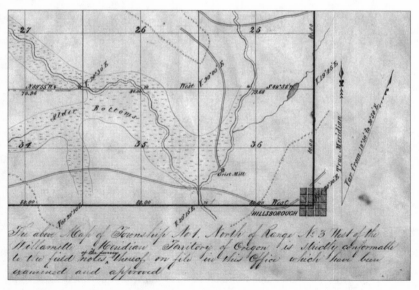

Figure 16. General Land Office Map: Township 1 North, Range 3 West, W. M.; Willamette Base Line through Hillsborough, Contract 2, William Ives, 1851. (USDI Bureau of Land Management, Portland, Oregon.)

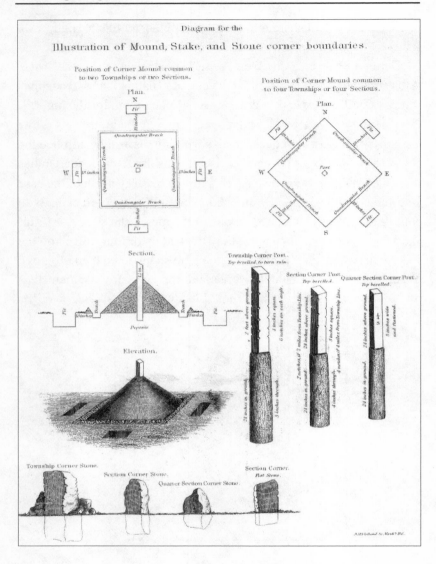

Figure 17. Mound, stake, and stone corner boundaries. (Reprinted from John M. Moore, *Instructions to the Surveyors General of the Public Lands for the Surveying Districts Established In and Since the Year 1850.* Washington, DC: A. O. P. Nicholson, Public Printer, 1855.)

The next morning, the hungry men backtracked to find Butler Ives. After breakfast, William Ives returned to the base line with extra men to pack food. Meanwhile, Butler Ives took the mules back to the Tualatin Plains for more supplies. After returning to the base camp, he and his companions packed food into the mountains, straining nearly a week under the heavy loads, before finally overtaking the company camped after six miles of hard travel.[52] On June 24, Butler Ives and McDermott relayed supplies to the survey party. In the first all-day ordeal, they made six miles in the mountains before camping for the night. June 25, they drove four hard miles in the rain. The next day they climbed three and one-half miles in a piercing rain before Robert Brevoort stumbled back to meet them.[53]

On the line ahead, William Ives strained through burnt timber and thickets that ripped his face and snagged his buckskin clothing. The surveyors willingly sweltered in the tough hide garments in summer and shivered in them in winter, for the buckskin clothes offered good protection from the ensnaring brush. Pushing forward, Ives and his men slid into deep ravines and climbed ridges that reached 2,000 feet above sea level. The deputy surveyor "passed what appeared to be highest Mountains, but could not see the Ocean on account of Fog and clouds . . . It was also as bad on each side of the line as far as I could see." On June 25, Ives and his men hunted over a half-mile without finding a place level enough to sleep.[54] "These Mountains are so steep that it is difficult to chain them with much accuracy," the deputy observed.[55] Peering through the dripping mist, Ives could make out the "Fir Hemlock . . . & spruce trees" that carpeted the slopes. Using names common at the time, the surveyors identified what is now called Douglas fir as "fir" or "red fir," present-day Ponderosa pine as "yellow pine," Oregon ash as "ash," and Garry oak or Oregon white oak as "white oak."[56]

On June 27, Ives stood on Hembree Ridge overlooking the Wilson River canyon where today Highway 6 winds between the interior valley near Banks, Oregon, and the city of Tillamook. Beyond the Wilson lay endless ridges of the Coast Range. One mile into Township

7 West and thirty-seven miles west of the initial point, William Ives ended the base line survey, stymied by the "steep rocky mountains" and the "impossibility of getting provisions carried along sufficient to last the surveying party through to the Pacific ocean and back" (Figure 1).[57]

Struggling a mile farther through the mountains, Butler Ives and Brevoort finally overtook the company. The exhausted men rested before hiking the five or six miles back to camp.[58] All hands trudged eastward for the next three days, camping near donation land claims where kind settlers gave them milk, vegetables, and fruit. From camp outside Portland on July 1, William Ives wrote John Preston, "I have this day returned from the line I started for the Pacific ocean without accomplishing the object . . . It is the worst 17 miles that I ever surveyed . . . If it is not very necessary that this line should be continued to the Pacific I must respectfully ask to be relieved from that part of my contract."[59] He assured the surveyor general that he and the men had "escaped the wild beasts of the mountains," adding that they had not seen "a mark of a human being west of the plains and no signs of animals larger than squirrels after getting ten miles west." Taking a positive tone, Ives closed, "The weather is fine now & I look for good country."[60]

Willamette Meridian South, June 9 to July 8, Seventy-Eight Miles

On June 7, in Oregon City, James Freeman's axemen Alonzo Stafford, William Humble, Elias South, and Joseph Barston signed their oaths. Freeman filled out his company two days later near the intersection of the meridian and base line. Joseph Hunt and Frank Brytone joined his crew as did Allan Seymour and John Stephenson, both of whom Freeman had met aboard the *Empire City*. The following day, twenty-nine-year-old Molalla schoolteacher William A. Starkweather signed on as a chainman. Alonzo Stafford, Zenas Moody, and Israel Mitchell also joined Freeman's crew. Mitchell, whose donation land claim extended into Multnomah, Washington, and Clackamas counties, had substantial surveying skills.[61]

Freeman looked forward to having US Geologist Dr. John Evans join the survey party. Evans planned to travel the meridian line and examine the country's geological characteristics using a barometer to record mountain elevations over which the line ran. Surveyor General Preston directed Freeman to share information with Evans and, if the geologist wished to camp with him, to "please permit him to do so."[62]

As of June 10, however, Dr. Evans had not yet reached the Territory. James Freeman left the initial point on June 11, traveling south on the Willamette Meridian under persistent cloudy skies that prevented use of the solar compass. Freeman and his men sat out June 12 and, by late the next day had advanced only two miles. When the clouds finally lifted, the company chained south into the Willamette Valley where mixed forest and prairie lands thirty miles wide and 150 miles long stretched from the Columbia River on the north to the Calapooia Mountains on the south. The northern area of the valley contained the Territory's largest communities: Oregon City, Linn City, Milwaukie, and Portland. To the south, Salem, Independence, Marysville (present-day Corvallis), Albany, and Eugene City marked ferry landings and other strategic sites along the river.

Crossing the Willamette River, Freeman surveyed the meridian across French Prairie east of where Wilsonville now spills south across the river. Farm families had widely settled these broad grasslands before 1850, building their dwellings and barns on high ground near woodlands that bordered their fields. Where golfers now lope the greens of a semiprivate course, Freeman saw rich soil and abundant, large-diameter fir timber. He observed presciently that the rolling country between the base line and the first standard parallel thirty miles to the south "is fast settling . . . there appears to be none that will not be eventually settled."[63]

Freeman crossed the twisting Pudding River twice. At the meridian's eighteenth mile, on June 20, the company camped in the bottomlands east of the river, where the surveyor wrote Preston, "The settlers [say] that the meridian line will intersect the Cascade Mountains leaving the Willamette Valley and extending far east of Umpqua passing through an unbroken chain of mountains." Freeman expected difficulty chaining the meridian through the mountains, "even if the public interest could be benefited."[64]

Between June 21 and 27, Freeman surveyed the meridian south twenty-four miles through four townships. He pushed his crew hard. The men were on the line at sun-up. Regardless of the terrain or weather conditions, they worked until noon, stopping for a midday meal, and persisted through the afternoon, pushing the line forward until they lost the sun, or even longer if they used the magnetic compass (Figure 18). At

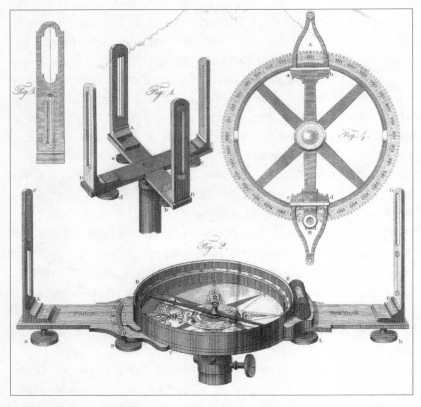

Figure 18. Surveyor's Compass. (Reprinted from Charles Davies, *Elements of Surveying and Navigation, with Description of the Instruments and the Necessary Tables.* New York, New York: A. S. Barnes & Co., 1854.)

dusk, the company ate supper and then slept, awakening the next morning at dawn to begin again. Unaccustomed to such intense physical labor, John Stephenson, one of the young men who accompanied Preston to Oregon, collapsed from exhaustion about two weeks out. After a few days of rest, however, Stephenson recovered and rejoined the crew.[65]

Moving south between present-day Woodburn and Molalla, Freeman crossed Bear Creek and Butte Creek. Fir, pine, and hemlock forests blanketed the foothills east of the meridian; white oak, alder, and willow grew along the streams to the west. Mile after mile of grass-covered prairies marked with stands of oak and fir stretched out along the line. Through country now marked by scattered rural development, the deputy surveyor passed John W. Newman's 600-acre claim straddling the Clackamas and Marion county line, as well as Jesse Choate's 620-acre Clackamas County claim. One day, while calculating the position of distant high peaks in the Cascades for the geodetic notes, Freeman broke a transit.[66] With the valuable surveying instrument now useless, cumbersome baggage, Freeman left it with a nearby settler. William Glover or perhaps Morgan Rudolph, through whose fields the meridian ran, agreed to keep the transit until the deputy surveyor could return for it (Figure 19).

Freeman surveyed the line over Abiqua Creek, its left-bank bluff twenty-five-feet high, pushing across open prairies and low hills of the Cascades. In Township 7 South, the company crossed Drift Creek and climbed into the hills where groves of white oak studded the slopes east of the broad Willamette Valley. Crossing the North Santiam River, the company climbed higher into the foothills southeast of present-day Stayton.[67] On July 2, wading "Crabtree fork," Freeman passed east of David Claypool's fields on Beaver Creek to mark the Meridian's 66th mile.[68] Over the next six days, the company crossed Hamilton and McDowell creeks in the hills east of present-day Lebanon and skirted the westerly reach of Marks Ridge to ford the South Santiam River north of present-day Sweet Home, Oregon. Freeman ran the meridian line south four miles into Township 14 South and stopped

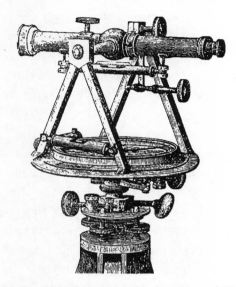

Figure 19. Transit. (Reproduced by permission from Francois D. Uzes, *Illustrated Price Guide to Antique Surveying Instruments and Books.* Rancho Cordova, California: Landmark Enterprises, 1980.)

on the north bank of the Calapooia River. The Willamette Valley farmers had advised him accurately. Deep blue canyons and steep green ridges rolled ahead as far as he could see. Directed to run the meridian close to settled areas, the surveyor now confronted isolated and impassable country.

A month into their contracts, William Ives and James Freeman had yet to survey more than half of their lines. William Ives still had to complete the eastern segment of the base line from Portland to the Cascades and most of the meridian north of the Columbia River. Freeman faced the toughest portion of his meridian survey. Cloudy weather had hampered both men, delaying Freeman's start by a week and forcing Ives to temporarily abandon the meridian and to survey the base line instead. Behind schedule, Ives prepared for the next leg of his expedition while Freeman looked out over the Calapooia and weighed his options.

Chapter 3 – Summer, 1851

Meridian

"Getting along as well as the country will admit."

As his deputy surveyors pushed hard on their lines, Surveyor General Preston plowed through the blue bundles stacked on his desk. After the *Oregon Spectator* published the entire Donation Land Act text on April 25, 1851, a barrage of letters painstakingly penned on thin paper had poured into his office from anxious settlers confused about the Act's provisions. Dreading the mail's arrival and hopelessly behind in responding, Preston used the newspaper to advise settlers on the surveys' progress and to answer questions about such matters as claim occupancy, claim portions for wives, and allotments to survivors following a claimant's death.[1]

Putting his paperwork on hold, Preston packed for a three-week trip to the Umpqua Valley.[2] With Nathaniel Coe, newly named as Oregon's postal agent, the surveyor general boarded the "iron propeller" steamer *Willamette* for Portland. From there, the two men descended the Columbia River and then steamed down the Oregon coastline to the mouth of the Umpqua River where they arrived on July 8. Preston, who hoped to confer with local officials, and Coe, who intended to appoint postmasters for newly-created Umpqua County, stopped first at the town of Gardiner on the north bank of the river. Walking along the shore of the estuary, the two men peered at the cedar plank lodges of the lower Umpqua clustered along the bank. Observing the rectangular shelters built over a pit in the earth, smoke rising from their plank roofs, Preston — no anthropologist — assessed the residents as a "tribe of coast Indians who mostly reside under ground." After a special seafood chowder feast in his honor at Umpqua City, Preston took a steamer inland to Scottsburg, where he called on Stephen Chadwick, with whom he had traveled to Oregon.[3]

The surveyor general left Scottsburg and traveled farther inland to Jesse Applegate's farm at the base of Mount Yoncalla. A member of the Oregon Trail "Great Migration" of 1843, a representative to the territorial legislature from Yamhill County in 1845 and from Polk County in 1849, Applegate was also a skilled surveyor. Well-read, independent, and energetic, he influenced territorial politics through wide correspondence and by inviting prominent visitors to his farm. The two men discussed Oregon politics, and then Applegate brought the conversation around to the public land surveys. When his host hinted at wanting a survey appointment, Preston advised him that possession and experience with the solar compass were "indispensable qualifications for a deputy." Never having owned or even seen Burt's compass, Applegate abandoned his bid for employment.[4]

Base Line East, July 2 to July 12, Thirty-Six Miles

In Portland, a frustrated William Ives hunted for qualified assistants. "I have not as many men as I want to make a full company," he wrote Preston, "but I think that I shall have sufficient to keep the work a going if I do the hardest of it myself as I have so far except what my brother has done."[5] He hired James du Moulin and William Gilcrist to replace Nicholas Coarser and James Swinhart, and signed on eighteen-year-old Samuel Gatton, an Ohio native who had arrived in Oregon the previous November. Gatton lived with his grandparents on Swan Island, now a peninsula in the Willamette River northwest of Portland's downtown core.[6]

On July 2, Ives resumed work east of the Willamette River, plunging through the wetlands past John Carothers' log house to survey the base line along what is now Stark Street in east Portland. Five years prior to the base line survey, fire had charred much of the area of today's east Portland; consequently, William and Butler Ives had to crawl through thick brush and fallen trees.[7] One day's travel east of the city, the country turned dry and desolate. On July 4, William Ives sent his younger brother in search of water. Butler Ives returned to the Portland-Vancouver road and then traveled up the Columbia. Obtaining water along the way, he went on to the Sandy River, where he planned to meet William Ives and the rest of the company. William Ives, meanwhile, continued east across Beaver Creek and soon joined his brother at the Sandy. One weary, thirsty crew member quit at the river, though he did agree to take the pack mules — useless in the rugged country ahead — back with him to a farmer's pasture (Figure 20).[8]

Leaving Sandy River on July 7, the weary men strapped packs onto their backs and edged up the Cascades through dense fir and hemlock forests. As William Ives and the company pushed uphill, Butler Ives followed three miles behind, crawling forward over downed logs to relay supplies to the party.[9] For three days, William Ives and his men cut through thickets to cross the south slope of Larch Mountain. They climbed higher into the mountains, sweating as they scrambled up the thickly timbered ridge above the Bull Run River. There, the surveyor reached what he thought to be the crest of the Cascades. Preston's map showed no definite summit. The mountains, he declared, "rise somewhat gradually . . . so that one passes almost insensibly into [them] unless particular notice is taken."[10] Four miles southeast of Larch Mountain, on July 12, Ives marked the east end of the base line on the boundary between Ranges 6 and 7 East with a conical stone monument that measured four feet in diameter at its base and stood three feet high (Figure 1). As William Ives and his men piled the last rocks on the sturdy pier, Butler Ives and McDermott left the Sandy River to return to the meridian line north of the Columbia River. They rounded up the grazing mules and packed down the road toward Vancouver. At dusk, insects swarmed on the animals' oozing eyes and raised blistering welts on the men's exposed skin.

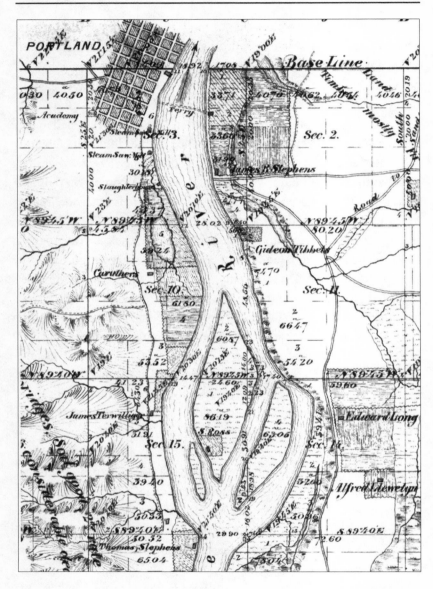

Figure 20. General Land Office Map: Township 1 South, Range 1 East, W. M.; Willamette Base Line to Sandy River, Contract 2, William Ives, 1851. (USDI Bureau of Land Management, Portland, Oregon.)

"The mousquitoes" said Butler Ives, whose spelling rivaled that of Captain William Clark, his predecessor in the Oregon Country by fifty years, "trouble us some."[11]

Willamette Meridian South, July 9 to August 11, 124½ Miles

On July 9, James Freeman gazed over rock ridges, some as high as 4,000 feet. "On crossing [Calapooia River] at this point," he observed, "the line would extend into the Cascade Mountains through which it appears impracticable to extend the Meridian Line."[12] The situation warranted making an offset, a line measured at right angles from the true meridian to avoid the impassable terrain. Although an offset would inevitably cost him time, Freeman decided on the lower-elevation route and backtracked four miles from the river to head west.[13] Extending from present-day Sweet Home to Halsey, Oregon, the eighteen-mile-long offset avoided the Calapooia River canyon, the rugged Mohawk and McKenzie gorges, and the mountains dividing the Middle Fork Willamette River and the Row River.

Moving west along the offset line, the men passed through miles of oak savanna. "On the Calapooyea Creek," Freeman observed, "is a fine settlement & good farming land north of the creek level or rolling the largest part of which is fir & oak openings on prairie."[14] Named for the Kalapooian people — as is Calapooya Creek in Douglas County, although spelled differently — the stream Freeman encountered on the offset rises in the western slopes of the Cascade Range and enters the Willamette Valley at Albany, Oregon.

Through Township 14 South, Range 2 West near present-day Brownsville, Oregon, the company passed through open, low-lying prairies. As in the greater Willamette Valley, winter rains that seasonally turned the Calapooia prairies to swamps, sent farmers to better drained soils in the foothills. The grass-rich, clay loam prairies would long remain undeveloped, for as historian Peter Boag points out, "settlers had neither adequate technology nor a large enough labor force to check floods and drain the prairies for cultivation." Not until the 1870s would farmers drain the prairie wetlands by channeling between sloughs, the "running water [cutting] deeper until the sloughs and lakes became a connected stream" and emptied out. Under this extensive "ditching," the streams that "had no definite channels but spread out over the floor of the valley." disappeared along with the ducks, geese, and muskrats dependent on the prairie for survival.[15] Turned first into hay fields and house lots, the prairies would, after the introduction of irrigation systems and flood control, eventually give way to urban expansion.

For now, however, James Freeman saw only agricultural promise in the oak-studded countryside. He asked farmers along the line for information about the location, number, and occupants of nearby land claims. "On the right side of the [river] bank [in Township 14 South]" Freeman's contacts informed him, "reside . . . Stewart Lewis, Thomas Fields, Anna [Splawn] Levi Field, C. Finley & R. Montgomery And on the left side, Wm. Robnett, Thomas Woodfin, Wm. McCaw, Robt. & Wm. Glass & Thomas Riggs."[16] The Stewart claim, the deputy noted, "is the most eastern claim on Calapooyea & extends to the meridian line . . . Finley's flouring mill is in T 14 R 2. Kirk's house is on the n.e. qr. of Sec. 1 T 14 R 3 W.M. [south of present-day Brownsville] where the Eastern road from Oregon City crosses the Calapooyea Creek" (Figure 21).[17] Lewis and Elizabeth Stewart farmed about three-quarters of a mile south of the point at which Freeman began the offset near Sweet Home. Tennessee-born Richard Finley and his family had settled on the river in 1847 and built a grist mill the following year. Finley's wife's parents, Alexander and Sarah Kirk, had settled their claim in the Calapooia River Valley in October, 1846.[18]

On July 13, Freeman left the prairies of the Calapooia and entered the main Willamette Valley, its broad expanse framed on the western horizon by the Coast Range. Ending the east-west detour on the boundary just north of Halsey, Oregon, Freeman turned south again on the offset meridian across rich, level prairie bordered with timber. In camp that evening, he recopied his field notes for seventy-eight meridian miles, intending to ship them to the surveyor general.[19]

As it happened, Preston, who was on his way back to Oregon City from Jesse Applegate's, encountered Freeman east of Muddy

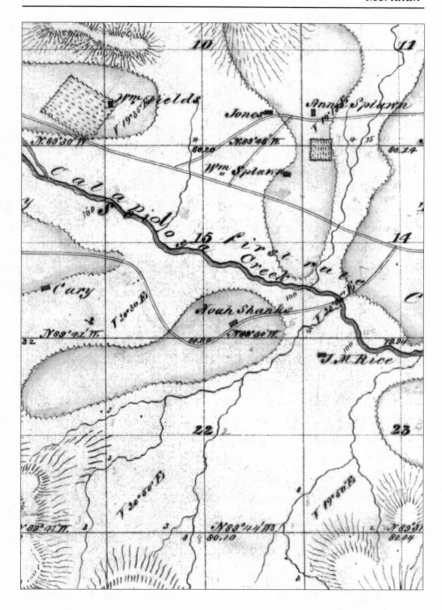

Figure 21. General Land Office Map: Township 14 South, Range 1 West, W. M.; Willamette Meridian, Calapooya Creek, Contract 1, James E. Freeman, 1851. (USDI Bureau of Land Management, Portland, Oregon.)

Creek and Prairie Precinct, near present-day Harrisburg, and collected the notes in person. In Oregon City a few days later, Preston wrote Commissioner Butterfield to explain the reasons for Freeman's offset, adding that his deputies "on the Base & Meridian lines progress slowly with their lines."[20]

James Freeman made good time through Township 16 South, southeast of present-day Junction City, on June 16, crossing several claims and plowed fields where Nancy Vaughn Ferguson, a widow from West Virginia, farmed the property with her oldest son, Eli Ferguson, and five younger children. The deputy surveyor bought provisions from William Tyler Vaughn, who operated a store out of his dwelling near the Willamette Valley's main road, not far from present-day Coburg, Oregon.[21] Known as the "west-side road" — a different road traversed the valley on the east side of the Willamette River — this important north-south route hugged the hills south of Marysville before crossing the Long Tom River and turning to the southeast. South of Eugene City, an extension of this route formed the territorial road to the Umpqua and Rogue River Valleys.

The company's progress slowed as they chopped their way south through thick brush and timber. On July 20, after a grueling day, Freeman was forty-one miles out on the line and had averaged less than three miles a day. The company camped that night near Eugene Skinner's farm at what is now Eugene, Oregon. The settler, who had taken his claim in 1847, operated a ferry across the Willamette. From camp, Freeman wrote Preston, "The level open country continues until the last 6 miles in which we have crossed the Willamette 5 times . . . If we have good weather I have no doubt but we shall be at Oregon City [soon] as we will doubtless run 4 miles per day" (Figure 22).[22]

Freeman left the Willamette River and climbed quickly into the hills. In Township 19 he crossed Camas Swale Creek Valley and again hit the Cascades.[23] On July 24, the company came south on line about a mile west of the valley of the Coast Fork Willamette River. Scrambling up Hobart Butte near the south boundary of Township 22, Freeman set a township corner on the southerly slope and gazed out

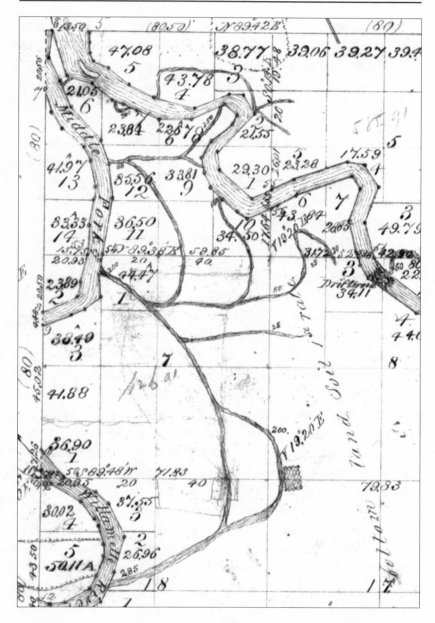

Figure 22. "We have crossed the Willamette River five times." General Land Office Map: Township 17 South, Range 3 West, W. M.; Contract 1, James E. Freeman, 1851. (USDI Bureau of Land Management, Portland, Oregon.)

on the rugged Calapooia Divide, its highest peaks rising to nearly 6,000 feet.[24] Seeing only high mountains far to the south and east, Freeman decided on a second offset.[25] Turning west between Townships 22 and 23 South, he descended the hills into Scotts Valley, where William J. Scott, son of prominent Oregon pioneer Levi Scott, had settled his land claim in May, 1850 (Figure 1).[26]

Freeman ended the six-mile-long offset between Ranges 4 and 5 West, about three miles east of Yoncalla, and moved south again through mountainous country. On August 3, the company passed Sim Oldham's cabin and forded the creek bearing the settler's name. Over the next week, the company made thirty-five miles to survey the meridian through seven townships. They crossed Calapooya Creek west of present-day Oakland, on August 4, to reach the ridge north of the stream.[27] Freeman and his men descended the hills the next day to the North Umpqua River at Whistler's Bend. Crossing the North Umpqua, they pressed south through the mountains east of present-day Roseburg, climbing across the steep easterly slope of Dodson Butte.

Over the next several days the company made slow headway through the rugged hills east of present-day Interstate 5 as it passes through Tri-City and Myrtle Creek, Oregon. On August 11, they ferried across the South Umpqua River. "The lands along the Umpqua," Freeman observed, "are unsurpassed in beauty and fertility of soil but between the streams are mountain ranges which will not be cultivated." Ascending these steep ridges, the surveyor and his men gazed out over the rugged Rogue-Umpqua Divide.[28]

On August 12, about two miles east of settler Joseph Knott's cabin, near present-day Canyonville and the Cow Creek Band of Indians' Seven Feathers Hotel and Casino Resort, an exhausted Freeman ended his survey. He stopped midsection, pounding in a quarter-section post on the west boundary of Section 30, Township 30 South between Ranges 4 and 5 West.[29] "From this point," he wrote Preston from camp that evening, it was "impracticable as well as useless to continue the line . . . a great part of [the country] . . . cannot be surveyed & if surveyed could not be settled" (Figure 1).[30]

Having surveyed 202.5 miles of meridian, Freeman and his weary company dropped down the steep slopes to the Oregon-California road. Heading north along the road, they met Jesse Applegate hunting stray horses near Yoncalla. Applegate invited the weary crew to recuperate at his home. The men accepted the invitation while Freeman continued on north. "Met Mr. Freeman on the mountain on his way to [Oregon] city," Applegate wrote John Preston a few days later. "Please inform Mr. Freeman that his young men have so far recovered as to be able to take up the line of march in the morning."[31]

Willamette Meridian North, July 18 to August 16, 91 Miles and 44 Chains

On July 13, as James Freeman chained south through the Willamette Valley, Butler Ives and William McDermott hiked north of the Columbia toward the point on the meridian that William Ives had abandoned a month earlier. Along the way, the men camped for two days on the west side of Vancouver Lake, where Ives trained his theodolite on distant peaks in the Cascades to record the prominent geographic features. Sending McDermott to Portland to deliver these calculations to his brother, Ives remained alone in camp. That night, the frantic, mosquito-ridden mules broke loose and ran. He corralled two of the animals the next morning, but when McDermott returned, the men circled for an hour without catching the other mule.[32]

Having decided to end his survey of the Willamette Base Line, William Ives returned to Oregon City with his men. Short on supplies and his credit stretched tight, he called on Preston. Disappointed to find the surveyor general not yet returned from the Umpqua country, Ives wrote, "When I return [from Puget Sound] I shall be owing mostly to my men about two thousand dollars ($2,000) . . . I have promised to pay them when I get the line done to Pugets sound and return to Portland which I am in hopes of doing by the 20th of August. If you can have the notes examined that I have sent into the Office so as to pay me for them on my return or have it ready by the 15th of August it will be a great help to me."[33]

On July 17, freshly provisioned, William Ives returned to the Willamette Meridian east of present-day Scappoose, Oregon, where his brother and McDermott waited. William Gilcrist remained with the company. Joseph H. Lambert and Friedrich Martens signed on as replacements for the survey to Puget Sound.[34]

William Ives started for the Kalapoodle, or Lewis, River, while Butler Ives came behind, leading two loaded pack animals over the swampy terrain. William Ives followed the Lake River for a distance, passing near present-day Ridgefield, Washington. Butler Ives overtook his brother and pressed ahead, nearly drowning the mules while swimming them across the sloughs. Nearly sick from the wet and cold, he reached the Lewis River near present-day Woodland, Washington, where he rested at Judge Lancaster's home. Columbia Lancaster (1803–1893), who had arrived in the Willamette Valley in 1847, served as an associate justice of the supreme court under the provisional government and was an unsuccessful Whig candidate for Territorial delegate to Congress in 1849. Lancaster settled a donation land claim at the mouth of the Lewis River and, when Ives visited him, represented Vancouver County on Oregon's Territorial Council.[35]

The next day, Butler Ives waited for McDermott, who had stayed behind to pasture a mule at Vancouver. Meanwhile, William Ives and his men crossed the Lewis to chain four miles of the meridian on July 20.[36] In the hills north of the river, Douglas fir, western hemlock, and grand fir blanketed the ridges. Rhododendron, sword fern, chinkapin, and Oregon grape choked the forest floor. As they pushed toward the Kalama River, the men crawled for hours through burned timber, downed logs, and brush so dense that they could not get the pack animals through.[37]

As his brother edged northward on the meridian, Butler Ives and three men rowed a supply-laden boat three miles up the Lewis River. The men hid the craft, strapped packs onto their backs, scrambled into the mountains, and overtook the company that night.[38] Having traveled thirty and one-half miles north of the base line on the south slope of the ridge between the Lewis and the Kalama, William Ives rose at dawn to write the surveyor general, "I am getting along as well as the country will admit . . . my mules have not been of any use to me for a month back. I have two of them at Calapoodle or Lewis River and one is taking care of himself because we could not catch him." From this point to the Cowlitz River, Ives advised Preston, the men must pack all their supplies on their backs. The country, he said, "lays hard ahead."[39] Closing his letter on a cordial note, he added, "If you should see fit to call on me at Puget's sound you will find my house open for accommodations."[40]

On July 22, before William Ives trekked north toward the Kalama River, he sent Butler Ives and Robert Brevoort down to St. Helens, Oregon, to mail his letter to Preston and buy supplies. The young men backtracked south and retrieved the hidden boat on the Lewis River. Heading downriver, they helped McDermott swim the mules over the river and sent him ahead to the Kalama. Ives and Brevoort stayed the night as Lancaster's guests. The men finished their errands at St. Helens, loaded both boats, and started on the return trip. Rowing hard up the Columbia against a stiff wind, they reached the Kalama at noon. They left one of the boats and some of the provisions with a settler and proceeded in the second boat. "The water in the river," Butler Ives observed, "is very clear . . . cool & filled with salmon and suckers."[41]

On July 24, Ives and Brevoort waited awhile, hoping to meet some of the survey party. When no one appeared, they continued upriver another two miles where, near the meridian's intersection with the Kalama, they saw cleared brush and blazed trees, certain evidence that the company had already come through. Leaving the boat hidden about a mile downriver from the line, Ives and Brevoort hoisted sixty-pound packs on their backs and climbed into the mountains. In camp that night near the meridian, they could see the glowing ashes of smoldering fires the company had set to burn brush.[42]

William Ives and his men descended the ridge between the Lewis and Kalama rivers' watersheds, crossing between Green Mountain and Devils Peak. On July 23, they crossed the Kalama, its swift-flowing water clear and cold.[43] Depending on the width and depth of streams, the surveyors crossed any way they could that would not jeopardize the equipment, field notes, and maps. The men looked for natural fords, sometimes searching far from the line for a crossing. On narrow, deep streams they cut saplings and dropped them across as a bridge. Most of the time, the men simply waded into the water. North of the Kalama, Ives scrambled up a rocky cliff, chaining the meridian through hills that reached 2,000 feet in elevation. On July 24, three miles north of the Kalama, he climbed the ridge between Goble Creek and North Fork Goble Creek and, the next day, camped in the hills roughly east of Monticello, present-day Longview, Washington.[44]

Butler Ives and Robert Brevoort chased the company five miles through the brush-choked hills before stopping at dark. That night, they could hear the voices of the men ahead in camp on North Fork Goble Creek.[45] The next morning they hiked one mile to overtake the survey party. Butler Ives' relief turned to dismay when he learned that two days earlier his brother had sent men to retrieve the boats and take supplies up the Cowlitz River. Although the two assistants would easily find the boat he had stored with the settler, Ives knew they could not find the craft hidden on the Kalama.

William Ives and his men strapped on loaded packs. Leaving North Fork of Goble Creek, they set the corner between Townships 7 and 8 North, chopping through thick briars that tore their skin as they plodded north toward the Coweeman River. Then, in an ordeal resembling a nineteenth-century cross-training exercise, Butler Ives hiked back six miles to the Kalama River to camp again that night without supper or a fire.[46] The next day, his second without food, Ives reclaimed the cached boat and rowed down the Kalama. He stopped at the farm where he had stored the other boat, only to learn that the men had already been there and had headed up to Wallace's. Victor M. Wallace, who had come to Oregon in 1847, settled on the east bank of the Cowlitz in 1850 and staked a 640-acre donation land claim. When Ives arrived, Wallace informed him that his crew had left the previous day with the mule string and instruments, headed for the Cowlitz Farm.[47]

Butler Ives followed the packers some twenty-five miles north along the trail on the east side of the Cowlitz before camping for the night, again without food. The next morning, he hiked six miles to Warbassport, formerly Cowlitz Landing, only to discover that the company had left there that morning.[48] In 1851, the settlement promoted by Edward Warbass on the Cowlitz River in the summer of 1850 offered lodging and food for travelers to Puget Sound. Just three weeks prior to Ives' visit, Northern Oregonians, as settlers residing north of the Columbia were known by 1850, had met at Warbass' to discuss separating from the Territory.[49] Frustrated by trying to conduct political and judicial business so far from Oregon City, dependent on canoe transportation between Monticello and Warbassport for travel and communication, and irritated by officials who ignored their needs, the settlers of North Oregon met several times during the summer of 1851 to craft a memorial to Congress.

After finally overtaking the pack team, Ives hiked up to the Cowlitz Farm to learn what he could about the country. Established in 1838, the Puget's Sound Agricultural Company operated Cowlitz Farm as a subsidiary to the Hudson's Bay Company, growing crops for English residents of the region while reinforcing Great Britain's claims to the country north of the Columbia River.[50] By the time the US government surveyors reached the Oregon Country, emigrants, who had found the best Willamette Valley land already taken, poured north to plant fields to wheat, oats, and barley and to graze thousands of head of livestock. England-born Company Agent George Roberts, who had arrived in the Pacific Northwest in 1831 and served at Fort Vancouver until 1846, now supervised operations at Cowlitz Farm (Figure 23).[51]

While his brother chased the pack team, William Ives and his men chained sixteen hard meridian miles toward the Toutle River. In Township 8 North, the company ran the line just below the summit of Smith Mountain, where they had a fine view across miles of timbered ridges. Leaving the Coweeman River, Ives surveyed the line across Meridian Mountain, cutting through high brush on the ridges surrounding the headwaters of Ostrander and Hemlock creeks. Three miles south of the Toutle, Ives sighted present-day Silver Lake about two miles to the west.[52]

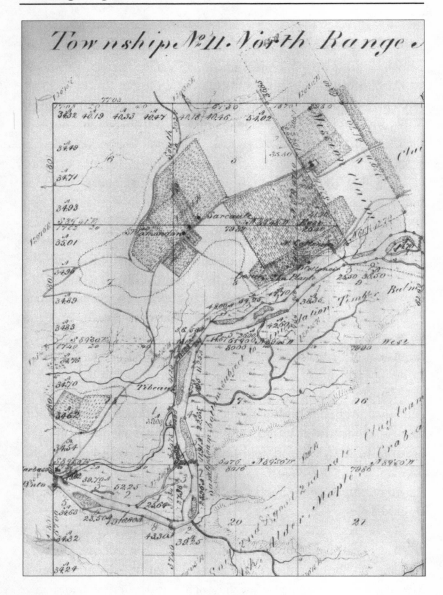

Figure 23. General Land Office Map: Township 11 North, Range 1 West, W. M.; Willamette Meridian, Cowlitz Prairie, Contract 2, William Ives, 1851. (USDI Bureau of Land Management, Portland, Oregon.)

On July 30, worried that his brother had not yet appeared, Butler Ives followed an Indian trail five miles up the Cowlitz River toward the meridian, threading notes on tree branches along the trail, should the company pass by. After waiting two days, Ives went down to Warbassport and found his brother at Warbass' Hotel. William Ives had lost two days after mistaking the Toutle River for the Cowlitz and had come out of the hills for provisions. "The map of the country through here is laid down wrong," Butler Ives observed with disgust, "the farms being placed some 15 miles too far south."[53] At the Cowlitz Farm the next day, the reunited company loaded the mules with fresh provisions and headed back southeast to the Toutle River to resume the survey. On August 3, William Ives started north on the line.[54] He crossed the fast-moving Toutle, chaining across Eden Valley and the Wilkes Hills to the southerly reach of the Cowlitz River plain. As they passed through the rolling country, the men stuffed themselves on huckleberries and salmonberries and sampled the fruit of the salal which they described as delicious.[55]

Butler Ives borrowed a horse from Agent Roberts at Cowlitz Farm and rode thirty miles up the trail towards Puget Sound to investigate the terrain ahead. He stayed that night with Joseph Borst, prominent Chehalis River Valley settler who occupied a land claim on Ford Prairie, and the next morning led the pack animals up the Cowlitz to rendezvous with his brother on the meridian.[56] HBC agent Roberts had estimated that about forty to fifty miles remained between the Cowlitz River, at the meridian's 65th mile, and Puget Sound. The first ten miles led across Cowlitz and Jackson prairies to the South Fork Newaukum River; the next fifteen to twenty miles of line ran through the hills flanking the North Fork Neweaukum River north to the Skookumchuck River. Between the Deschutes River, different than the river of the same name located in central Oregon, and the Sound, Roberts reported, the line crossed open country occupied by a number of settlers.

On August 7, Butler Ives helped the company pack provisions several miles into the hills north of the Cowlitz while McDermott returned to the Cowlitz Farm with the mules. The party crossed present-day

Highway 12 near its intersection with what is now Kiser Road, while pushing north to the South Fork Newaukum River. Assured that the survey party had enough food to last until they reached a settlement, Butler Ives departed for Oregon City. Knowing the surveyor general's eagerness to begin the township surveys, William Ives had promised Preston in his last letter, "If we do not see or hear from you my Brother will return . . . to run the Township lines that you spoke to him about at the Tualitin Plains."[57] Butler Ives stayed a night at Warbass' Hotel, canoed down to Victor Wallace's at the mouth of the Cowlitz, rowed two days up the Columbia to St. Helens, and eight miles up the Willamette Slough to camp for the night. On August 10, he rowed up the Willamette River to within four miles of Portland. In the city the following day by noon, he rested for a few hours before going on to Oregon City.[58]

William Ives camped north of the Middle Fork Newaukum River. Over the next six days he ran the line twenty miles through Townships 13, 14, and 15 North, crossing the North Fork Newaukum and the Skookumchuck River to reach the Deschutes River near present-day Highway 507, on August 13. At lower elevations, the meridian line now passed through fir and cedar and alder forests and, despite the sun's low declination and some sunless days, Ives had sufficient light to operate the solar compass.

Crossing the site of present-day Interstate 5 a short distance west of the Nisqually River, William Ives reached the southern shore of Puget Sound about ten miles northeast of Olympia, Washington, on August 16. Standing 110 miles and 44 chains north of the Willamette Base Line, he piled stones for a cairn on the shore of the Sound to mark the terminus, finishing his survey four days after James Freeman had driven the last post in the Umpqua country (see Figure 1).[59] "The north end of the Willamette Meridian," Ives wrote Preston, "is nearly west of the Hudson Bay Co. store house near Fort Nisqually and about three miles distant." [60] "Mount Rainier," he observed "cannot be plainly seen on account of smoke & clouds." Evoking a transitory scene, Ives added, "There are a number of Families of Indians around here. They are catching small flounders & some other fish & appear to live entirely on what they catch in the water and mud."[61]

Ives described the country north of the Columbia as burned over, much of it within the last five to ten years. The meridian from the Lewis River to the Coweeman River, he told Preston, had a "heavy original growth of timber that is mostly dead," while the timber from the meridian's 48th mile north to Puget Sound was "old growth & part young & middling growth" with heights from 100 to 250 feet and measuring from one to six feet in diameter.[62] The surveyor had seen only six donation land claims along the way: three on the 65th mile at the Cowlitz River, two west of the line near the 95th mile, and one between the Deschutes River and Puget Sound. "I am credibly informed," he told Preston, "that there is a settlement of five or six families on and near the mouth of the Nisqually River."[63]

Ives canoed southwest on Puget Sound to visit Olympia, officially platted the year before and, at the time of the surveyor's visit, a growing community containing "some 10 or 15 houses and situated on the fartherest point south of the sound & about 8 or 10 miles West of the Meridian."[64] With prominent access to Puget Sound from the Cowlitz Trail, Olympia — named port of entry for North Oregon in February 1851 — had developed rapidly as a center for commerce and government. His contract finished, Ives turned south from Olympia toward Oregon City.

Hiking for eight weeks in opposite directions from the initial point near Portland, William Ives and James Freeman had staked over three hundred miles of meridian and seventy-three miles of base line in Oregon Territory. These grueling surveys complete, the men would now embark on the even more exhausting and exacting challenge of laying out Oregon's grid of townships. Competitors for the upcoming contracts, Ives and Freeman hurried back to tackle Preston's diagram.

Chapter 4 – Fall, 1851

Township and Section
in the Willamette Valley

"Musquitos thick and watch dogs plenty."

Surveyor General Preston grappled with the Donation Land Act. Bewildered by the document's complex provisions, settlers sought his advice, writing in astonishing detail of their widowhood, citizenship status, and other personal circumstances. Pressure on the surveyor general to interpret the Donation Act intensified when ambitious young journalist and attorney Asahel Bush demanded clarification of specific sections. Bush, who had arrived in Oregon in 1850, became editor and publisher of the *Oregon Statesman* — initially located in Oregon City — in 1851 and had rapidly gained political power. That summer, Bush badgered Preston to clarify the Act's fourth section.[1] "Are persons of color excluded from the Donation Grant by it," he inquired, "or are all admitted to its privileges, without regard to color?"[2] Persons of color, it turned out, were not allowed to claim land.

In 1851, the Oregon Territory simmered with the tensions that would erupt a decade later in the bitter conflict of the Civil War. As historian Jeff LaLande has observed, Oregonians carried west the "cultural baggage of sectional allegiance, racial prejudice, and political philosophy," a heritage that fostered intense debate about the place of blacks in the Territory.[3] "Many Oregonians detested slavery, but not on moral grounds," writes LaLande. "Oregon settlers were troubled by the split over slavery that bedeviled their fellow Americans back east — and by the specter of blacks, free or slave, as economic competitors and social intruders. In reaction, Oregonians repeatedly produced constitutions forbidding both slavery and the residence of 'free Negroes' within Oregon."[4] Now, Bush queried the surveyor general on the Donation Land Act's provision that land should be granted to "every white settler or occupant of the public lands, American half-breed Indians included, above the age of eighteen years, being a citizen of the United States."[5] "Mr. Preston . . . and myself," Judge Thomas Nelson observed, "have discussions almost daily upon the true interpretation of the land bill of the territory and hardly is one question disposed of before like Hydra heads others spring up in its place."[6]

When he wasn't wrestling with the Land Act, Preston crafted a plan for surveying the Territory's western valleys. The surveyor general was impatient to get the surveys underway now that Butler Ives was back in Oregon City, and William Ives and James Freeman were due to arrive shortly. Preston chose townships in the oldest settled areas for survey in fiscal year 1851–1852 and another group for the period ending June 30, 1853. In the next two years, Preston planned to complete the survey of the densely settled Willamette Valley from Oregon City south to Albany. He used the *Spectator* to publicly announce the impending commencement of the public land surveys, admonishing settlers who had not yet had their donation land claim boundaries marked to notify him promptly of the precise tracts of land they claimed.[7] The most skilled deputies would survey township boundaries — lines that formed a critical framework — while the less experienced surveyors would subdivide the townships into sections.[8]

∽

As clean as his primitive camp conditions allowed, Butler Ives kept his August 15 appointment with the surveyor general, who named him a US Deputy Surveyor for Oregon and offered him a survey contract for township boundaries. Ives borrowed $100 from Preston and took his brother's boat downriver to outfit his first solo contract in the Oregon Country. Stopping briefly at Milwaukie to order new legs for his tripod, he rowed on to Portland and rented a room at the Columbian Hotel at Front and Washington streets. From his window, Ives looked out over the busy commercial district along the Willamette water-front. Wooden wharves and warehouses lined the riverfront and two-story frame shops, hotels, and saloons crowded the adjacent streets. Ives bought provisions on credit at Northrup and Simonds and hired Thomas Spring to work for him at $60 a month. Appalled when the Columbian's proprietor charged him $4.00 for the first day's lodging, the frugal surveyor packed his bag, boated across the Willamette, and set up camp.[9] Heading toward Fort Vancouver two days later, Ives stayed the night at Switzler's Landing where John Switzler operated a public ferry on the south bank of the Columbia River. Despite having a room with a feather bed, the surveyor got little rest, "the night was warm musquitoes thick & watch dogs plenty. . . had no sleep at all." The next morning, Ives crossed the Columbia to Fort Vancouver to buy a mule, returned to Portland to board the animal in a settler's pasture, and rowed back up the Willamette to camp again across from Oregon City.[10]

On August 23, Preston awarded Butler Ives Contract 3 for the first township survey in the Territory — sixty-six miles of boundaries for six townships lying east of the meridian in the upper Willamette River basin — at $18.00 per mile. The townships covered a six-mile-wide, thirty-mile-long swath from the Columbia River south to Butte Creek east of Mount Angel, in an area cut by the Willamette, Tualatin, Molalla, and Pudding rivers and encompassing present-day Milwaukie and Canby (Figure 24). Main Street House proprietor Sidney W. Moss and prominent trapper, legislator, and marshal Joseph L. Meek signed as Ives' sureties for $2,376.00.[11] One of three deputy surveyors in Oregon and the only one on hand at the time, Ives had no competition for the job. "Although contracts were supposed to be given to the lowest bidder," surveying historian C. Albert White has observed, "in

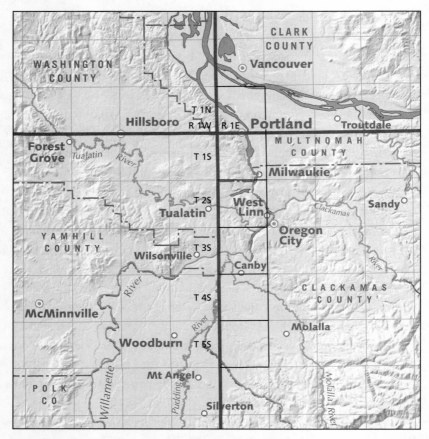

Figure 24. Survey Contract 3, Township 1 North, Range 1 East and Townships 1–5 South, Range 1 East, W. M.; Butler Ives, 1851. (Map image from Stuart Allan, et al., *Atlas of Oregon.* Eugene, Oregon: University of Oregon Press, 2001.)

fact there was no bidding process as such, the work was contracted at a stated price per mile."[12] The small number of available surveyors experienced with the solar compass further narrowed the initial field of qualified contractors to Butler and William Ives and James Freeman.

After a hard week's travel from Puget Sound, William Ives reached camp on August 26, where he sold his brother a solar compass for

$190, a mule for $75, and paid off his crew. Butler Ives immediately rehired three of the men to round out his own party of two chainmen, a flagman, and an axeman.[13] As the younger Ives and his men walked north from Oregon City the next day, US Geologist B. F. Shumard, Dr. John Evans' assistant, accompanied the group.[14] With "J. C. Clark, D. C. Raymond, W. Gilcrist, W. Lewis, and J. Pickring," Butler Ives began the survey at the corner of Township 1 North and 1 South, Ranges 1 and 2 East, in what is now east Portland.[15]

Chief Clerk John Moore's manual determined Ives' route for surveying townships according to their positions in relation to the meridian and base line. For Contract 3, the deputy followed the procedure outlined for townships lying north of the base line and east of the principal meridian. He began the intricate pattern in the southeast corner of Township 1 North Range 1 East, chaining north to measure the "random" (temporary) lines west and the "true" lines east, thereby accommodating variations in the earth's curvature, casting "the excess over or deficiency under four hundred and eighty chains on the *west end* of the line as required by law."[16]

As he extended the framework established by his brother and James Freeman on the Willamette Meridian, Butler Ives described the country in detail in his field notes and sketch maps. He noted trees including their species names, diameters, and positions in relation to the line. He listed prairies, swales and ridges, terrain, soil, rivers, creeks, lakes and ponds, towns, and roads. He also followed the General Land Office directive to record "Indian towns and wigwams. . . natural curiosities, interesting fossils . . . also all ancient works of arts, such as mounds, fortifications, embankments, ditches, or objects of like nature."[17]

Under typical Oregon clouds that hampered use of his solar compass, Ives surveyed the line in fits and starts two miles north toward the Columbia. Retracing his steps the first night, he and his men camped for the night at settler Perry Prettyman's farm. A Delaware native and physician, Prettyman had settled his 620-acre claim in 1849 in what is now the Mount Tabor vicinity of east Portland. Then, what began as a slow start turned into unrelenting disaster. The men worked in rain so hard they had to stop and seek shelter. On August 31, Ives made "2 miss tallys and [a] bad close" and spent the next three days backtracking to correct his errors. On September 3, the company camped at Lewis Love's farm in the Columbia bottomlands, where the New York native and his wife, Nancy, held a large land claim.

While at Love's, Ives fell ill with the malarial chills and shaking known at the time as "fever and ague." Introduced with small pox and other diseases to the Willamette Valley twenty years earlier, malaria had killed vast numbers of native peoples. One of Ives' men went to Portland for medicine, but, by September 6, the deputy was too weak to work. He and his men stayed west of the Willamette River at Albert Kelly's farm — Albert Kelly Park in Portland is named for this pioneer — where the kind Methodist preacher took Ives into his home.[18] Two days later his men took the feverish surveyor to Portland where William Ives met them to transport his brother by boat up to Oregon City. Near collapse, Butler Ives paid off all of his men but two, who went to work for William.[19] The surveyor lay ill for four days, his chill and fever worsening, in a house rented from Elizabeth Thurston, Samuel Thurston's widow. On September 15, he managed to sign his oath for finishing the first township in his contract.[20]

Butler Ives' hard-won notes and sketches for Township 1 North, Range 1 East, depict a rural landscape now vanished. The boundaries of the township enclose an area now inside the city of Portland, in a six-mile square roughly bordered on the east by SE/NE 39th Street and NE 42nd Street and on the north by an imaginary line running through the Columbia River and grazing the southerly tip of Hayden Island to cross Smith Island. The area's western boundary descends through today's North Portland, crosses the Willamette River near the Burlington Northern/Union railroad bridge, cuts through Forest Park, and touches the western border of Mount Calvary Cemetery where the southern boundary passes east through the Japanese Gardens, bisects downtown Portland and crosses the Willamette just north of the Morrison Bridge.

Where present-day Delta Park, Portland Meadows, and Portland International Airport extend between Columbia Boulevard and the Columbia River, Butler Ives found swamps. "The bottom lands of the [Columbia]," he observed, "are subject to an annual inundation lasting from 1 to 2 months in the summer season, leaving only a few of the highest places dry . . . they are badly cut up with bayous, stagnant ponds & lakes, which render nearly one half of them unfit for cultivation."[21] Where urban development now straddles the Willamette, settlers cultivated the fields on both sides of the Willamette River. Jacob Wheeler, James Thompson, and Lemuel Hendrickson farmed the rich soil east of the river, while William Blackston and George Kittredge worked the land to the west. Most of present-day northeast Portland remained undeveloped, marked only by the road from Portland to the Sandy River, now Sandy Boulevard, and the Portland-Vancouver trail. Where houses now pack the blocks between Fremont and Broadway, a half-mile-wide swath of burned timber stretched eastward for miles.

In late September, Butler Ives hired a new crew, including Wisconsin native George McFall, recently arrived in Oregon City from the California gold mines. Ives found an able, loyal assistant in McFall, who was digging potatoes on a claim east of town when the deputy hired him at $60 a month.[22] The surveyor went back to work on September 25, covering four miles the first day. McFall spent most of his time carrying a heavy pack — "transmogrified into a mule" as he described it. Over the next month, Ives surveyed new lines and remeasured others to correct old mistakes. On October 21, he returned to Oregon City and discharged his men, finishing the contract late, discouraged and very tired.[23]

As his brother recuperated, William Ives signed Contract 4 to survey a block of nine townships surrounding Salem, Oregon. The area encompassed the confluence of the Santiam and the Willamette rivers and stretched west to the oak-covered Eola Hills. Ives used part of the $3651 Preston paid him for the meridian and base line surveys to outfit his expedition. The money didn't go far. "The Deputy

Surveyors who run these lines," Preston wrote Commissioner Butterfield, "assure me that the compensation barely paid the actual expenses they incurred in doing the work."[24] Although a sick crew got Ives off to a shaky start — three of his men soon fell ill with ague — they recovered to make good time across Howell Prairie, west of present-day Silverton (Figure 25).[25]

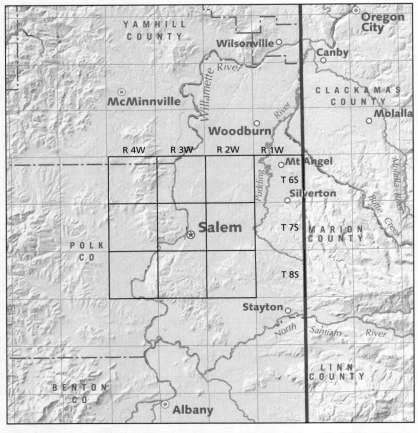

Figure 25. Survey Contract 4: Townships 6–8 South, Ranges 2–4 West, W. M.; William Ives, 1851. (Map image from Stuart Allan, et al., *Atlas of Oregon.* Eugene, Oregon: University of Oregon Press, 2001.)

Driving the surveys forward, Preston awarded James Freeman Contract 5 on September 17 to survey township boundaries. Three townships embraced the broad plain east of Salem; eight others encompassed the Santiam Valley, the North and South forks of the Santiam River and reached eastward to the Cascade foothills (Figure 26).[26] Needing more surveyors now that Butler Ives, James Freeman

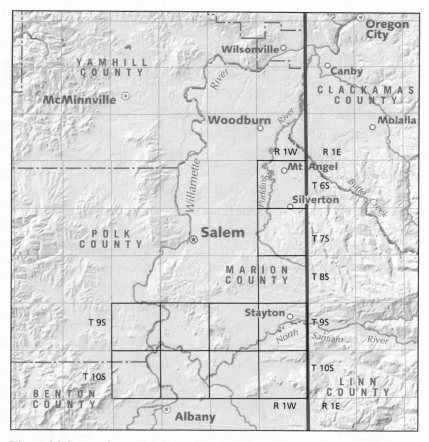

Figure 26. Survey Contract 5: Townships 6–8 South, Range 1 West and Townships 9–10 South, Ranges 1–4 West, W. M.; James E. Freeman, 1851. (Map image from Stuart Allan, et al., *Atlas of Oregon.* Eugene, Oregon: University of Oregon Press, 2001.)

and William Ives all had contracts, Preston hired his brother-in-law, George Hyde, as the Territory's fourth deputy surveyor. Hyde promptly resigned as head clerk in the office to apprentice briefly on Freeman's crew, joining chainmen Allan M. Seymour and Joseph Pownall, marker Zenas Moody, and axeman Kimball Webster.[27]

Webster, a twenty-two-year-old native of New Hampshire, had met Robert Elder the previous spring in the mines near Marysville, California, where Elder worked as a civil engineer. Although Webster, who worked for him, found the older engineer kind but eccentric, he welcomed Elder's invitation to join him in Oregon City. When Elder began the survey of the town lots on John McLoughlin's claim, in August, 1851, he again hired Webster as his assistant. This important project called for surveying the lots McLoughlin had sold prior to March 4, 1849. The results would enable Preston to certify title papers for the 195 lots and parts of lots settlers had filed in his office.[28] Now, in early October, Webster hurried south to join Freeman's party near Salem (Figure 27).

Webster proved a quick study at blazing bearing trees and hewing section posts, and, when the deputy surveyor noticed that his young assistant also excelled at plane trigonometry, he let him help with the calculations.[29] In late November, Freeman asked Webster to retrieve the transit broken on the meridian line survey the year before, pointing out on the sketch map the farm where the instrument was stored. Webster found the house without difficulty, paid the farmer for his assistance, and started back. By late afternoon, however, he realized that he could not possibly reach camp by nightfall and looked for a place to stay.[30] After walking until well after dark he saw a dim light in the distance. After stumbling into an icy slough in the dark, he finally reached the farmhouse and awoke the charitable occupants who gave him dinner and a bed before sending him on his way the next morning.[31]

Having heard news of Preston's search for additional surveyors, Jesse Applegate, who had been practicing with a solar compass, again asked the surveyor general for a contract. He would not have pressed

Figure 27. Kimball Webster, Oregon Deputy Surveyor. (Reprinted from Kimball Webster, *The Gold Seekers of '49*. Manchester, New Hampshire: Standard Book Company, 1917.)

for work, Applegate assured the surveyor general, "had I not supposed from your conversation at my house, it was not only convenient, but your pleasure to give me employment."[32] Whether for professional or political reasons, Preston turned Applegate down, hiring Joseph Hunt instead. Recommended by William Ives as having "served in every capacity of a surveying party from cook to a compassman," Hunt hired Daniel A. Thurston and George Plummer to work with him north of the base line in the vicinity of the Tualatin River and Hillsboro.[33]

Preston awarded George Hyde the first subdivision survey in Oregon. Hyde signed on New York native Timothy W. Davenport and two additional crew members to help subdivide five townships in the vicinity of the city of Salem, Oregon.[34] Hyde paced the township as he would a labyrinth, starting in the southeast corner, walking west one mile to the point between Sections 35 and 36 and then north six miles, and marking the quarter-section and section (one-mile) corners. At each mile, he headed east to the range line, placing temporary markers and noting total distance. Hyde next returned west, making offsets to the true line, correcting the temporary markers to reach the previously established monument, eventually reaching the township's north boundary.[35] In the weeks it took him to subdivide Townships 6 and 7 South, Range 1 West, Hyde walked over 250 miles, recording land features, timber, and soil types, and noting bearings for Benjamin Leonard's house, Mary Willard's barn, and other land claim buildings he could see from the line (Figure 28).[36]

As Hyde and the other deputies keyed these donation land claims to the survey lines, they changed the way Willamette Valley settlers related to the land. Previously, newcomers had used the natural landscape to define their claim boundaries in forms that were often irregular in configuration. Historian Peter G. Boag points out that prior to June, 1844, Oregon's provisional government permitted settlers to shape their claims as they chose. Land laws enacted in 1843 and 1844 called for square or oblong shaped claims. "Between 1844 and 1850 laws still allowed settlers to orient square or rectangular claims according to what the local landscape afforded."[37] While defining how claims could be acquired in the Oregon Country after December 1, 1850, Oregon's Donation Land Law prescribed claiming the landscape according to the rectangular survey. As Boag observes, "The taking of land in square form and in legal subdivisions, bounded by lines perpendicular and parallel to the (far distant) magnetic north, effectively terminated the older land-claiming process in which humans worked directly with topography."[38]

Back in Oregon City, Butler Ives took four late-October days to survey the boundaries of Dr. John McLoughlin's claim. This project, like Robert Elder's survey of the town lots on McLoughlin's property, figured in complex political maneuvers involving the former

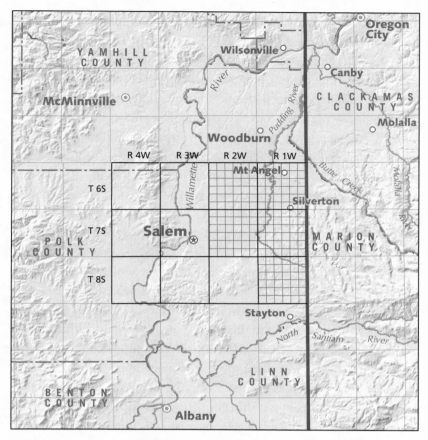

Figure 28. Survey Contract 6 (Subdivisions): Townships 6–7 South, Ranges 1–2 West and Township 8 South, Range 1 West, W. M.; George Hyde, 1851. (Map image from Stuart Allan, et al., *Atlas of Oregon*. Eugene, Oregon: University of Oregon Press, 2001.)

Hudson's Bay agent's long-disputed title to his land. Forced by Congress through the Donation Land Act to forfeit to the Territorial assembly portions of his land claim not given away or sold prior to March, 1849, McLoughlin waited as legislators and officials wrangled over his claim. Although Ives' survey could help the US government disenfranchise him, when Butler Ives signed his next contract, McLoughlin, whose generosity to the first American settlers in the

Oregon Country would earn him the title "Father of Oregon," signed the deputy's $8300 security bond.[39]

Preston issued contracts as fast as the surveyors could take them. He awarded Butler Ives Contract 8 in late October; on November 12, he sent William Ives northwest of Salem on Contract 9 to subdivide townships for which he had already run the exterior lines. Five days later, Preston assigned Joseph Hunt the subdivision of two townships in the Hillsboro vicinity, while giving James Freeman a contract southeast of Salem.[40] By the end of November, Preston had issued four contracts for boundary surveys of twenty-nine townships and five contracts to subdivide twenty-three townships. He shaded in the squares on his plan as his deputy surveyors turned in their maps. The steadily darkening diagram was clear evidence that the settled areas of Oregon, still a tiny fraction of the Territory's whole expanse, would in the foreseeable future all be defined by the linear survey.

In Oregon City, work piled up on the clerks' desks as Elder and his assistants transformed the deputy surveyors' field notes and sketch maps into an official record. After the clerks scrutinized the notes for accuracy, the draftsmen used the sketch maps to ink an original and two copies of the township plat.[41] The first map integrated information from the original township boundary survey. After the subdividing deputy had added his new information to this initial map and turned in his own field notes, the draftsmen crafted a final township map that contained all the critical information.[42] As the work mounted, Preston pleaded for more money. "Pay [for] additional clerks," he told the General Land Office Commissioner, "will be necessary to keep up the office work, recording field notes, proofs of settlement and cultivation of land claims." Advising Butterfield that the crushing work load required doubling the clerical staff, he added, "It is all important that all the office work be kept up with the surveys and proofs of donation rights, as every settler is anxious to receive his Patent in order to divide and sell."[43]

In October, 1851, Surveyor General Preston prepared his annual report for Commissioner Butterfield in Washington, summarizing the surveys accomplished to date and recommending new ones. Preston projected costs for the Oregon surveys for fiscal year 1852–1853, estimating $6,500 for his and the clerk's salaries, $4,500 for additional clerks in the office, and $2,000 for rent, fuel, and books. He predicted $60,000 in costs for surveying standard parallels, township, and sub-division lines for sixty townships, estimated at 5,000 miles of lines, at a rate not to exceed $12.00 per mile. Justifying the $73,000 budget, the surveyor general blamed the gold mines for keeping wages and product costs well above the prices paid in the East.[44] Bold on his own behalf, Preston complained to Butterfield, "The salary of the Surveyor General is entirely inadequate to the expense incurred in living here . . . It is as expensive for a family to live in Oregon City as in San Francisco. I trust that the Department will see the propriety of recommending that the salary of the Surveyor General is increased to meet the extraordinary expenses that have to be incurred in living in this territory."[45]

⤳

Near the end of 1851, Preston dodged battles between Whigs and Democrats in Oregon's increasingly stormy political environment. At the national level, basic philosophical and socioeconomic differences divided the two parties. Although conditions varied according to geographical location, Whigs typically supported commerce, educational reform, and temperance, and, while not against slavery, Whigs attracted abolitionists repelled by the more pro-slavery Democratic Party. Historically drawing their support from farmers, Southern slave states, and urban — largely Irish — immigrants, the Democrats supported expansion and slavery while denouncing corporate greed and domination by the social and economic elite.

Although these differences surfaced to some degree between Oregon's Whigs and Democrats — Oregonians tended to deeply resent the "imported" officials sent to them from Washington, DC — the most heated battles were fought over local offices, political patronage, and struggles between cliques of men pursuing power.[46] The Territory's government stood physically divided. Oregon City and

Salem each claimed to be the capital, with Whig officials centered in the former and Democrats in the latter city. On December 1, the Legislative Assembly convened in Salem and named it as the capital. Supreme Court Justice O. C. Pratt, a Democrat, went to Salem while justices Thomas Nelson, William Strong, and a few other legislators who refused to recognize the new capital, stayed in Oregon City with Governor Gaines.[47] Surveyor General Preston, a Whig, remained in Oregon City where he awaited word from Washington, turned an ear to Salem, and kept an eye on Asahel Bush's newspaper, still published in the former capital.

On December 23, the *Oregon Statesman* published a letter credited to a writer from the "Molalla Settlement" but most certainly crafted by Democrat editor Bush to harass the surveyor general. The correspondent calls erroneous parts of Preston's letter published the previous August advising each donation land claimant to "have all his corners established and his lines marked . . . in advance of the regular survey of the country."[48] The writer quotes Preston's statement that "within three months after the survey has been made, or where the survey has been made before the settlement commenced . . . each of said settlers shall notify the Surveyor General . . . of the *precise tract* or *tracts* claimed by them respectively under this law."[49] The "Molalla citizen" continues quoting Preston's stipulation that:

> In order to describe the precise *tract* or *tracts* claimed . . . the boundaries should be marked to enable the Deputy *Surveyors* to locate the lines of each claim properly as *he* [Bush's italics], progresses with the survey; if this is done it will prevent much confusion and delay in adjusting boundaries and locating claims on the township plats.[50]

"That this sentence is erroneously published, both from its errors to grammar and the construction of the Land Law must be evident to the most careless reader," snorted Bush. Calling it "a matter of surprise that the Surveyor General did not himself have it corrected," the *Statesman* editor added sarcastically, "But, as that respected functionary must be now continually employed in examining the notes of returning deputies, and dispatching others to the field, the letter has perhaps never fell under his notice."[51] Bush pressed on, "To make the sentence conform to the rules of grammar as well as the Land Law it should read:

> In order to describe precise tract or tracts claimed . . . the boundary lines should be marked by the Deputy surveyor as he progresses with the survey, in order to enable the claimant of each claim properly to locate it. If this is done it will prevent much confusion etc. [52]

Bush next focused his attack directly at Preston:

> As it now stands the Surveyor General appears to pay as little respects to the rights of the people under the law of Congress as he does to the law of language: he requires claims to be located *before* the survey when the law allows the claimant three months after [;] he requires *claimants* to make lines and establish corners which can only lawfully be done by deputy surveyors acting under a special contract and commission from himself . . .
>
> Not being able to obtain the services of the authorized deputies who came with Mr. Preston, the people . . . are employing men . . . whose ignorance and old fashioned instruments are held by him in proper contempt, and even if by accident they make or are qualified to make a correct survey, without his authority to do so, their work all goes for nothing.[53]

As one of the "imported" Whig officials sent to Oregon by President Fillmore and as a close associate of Governor Gaines, the surveyor general made a convenient target for the editor Bush's barbs. Despite solid progress on the public land survey, Preston now had Asahel Bush's attention and that promised trouble.

⟿

In December, Butler Ives subdivided Township 1 North, Range 1 East into sections, repeatedly crossing the Willamette and Columbia rivers in the southwest quarter of the township and fighting wind and rain to run his lines (Figure 29). He finished the township on December 15 and rowed hard two days up the Willamette to Oregon City, deciding to keep "a 'bachelor's den' for a month or two."[54] William Ives, on the other hand, kept on working in the rain. From north of Salem, where he and his men subdivided a township the week before Christmas, he wrote the surveyor general, "I have been able thus far

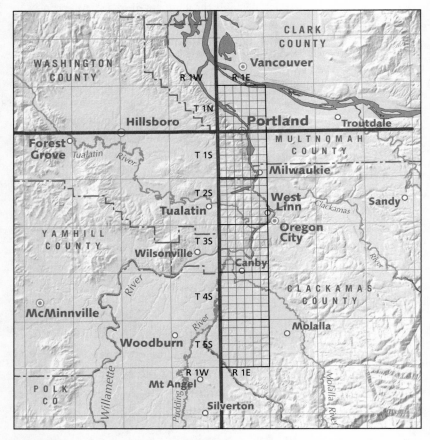

Figure 29. Survey Contract 8 (Subdivisions): Township 1 North, Range 1 East and Townships 1–5 South, Range 1 East, W. M.; Butler Ives, 1851. (Map image from Stuart Allan, et al., *Atlas of Oregon.* Eugene, Oregon: University of Oregon Press, 2001.)

to run about half of my work by the sun and most of my closes have been very good but occasionally a wild one."[55]

Misery and costs went up in bad weather. The relentless winter rains soaked the surveyors' clothing and bedding, forcing them to either seek shelter for which they paid settlers' fair prices, or to rent housing in Oregon City. Six months into their assignments, the rush of contracts was just now fully underway. On Christmas, 1851, "a dull, lonesome day," James Freeman, William and Butler Ives, and Joseph Hunt stood mid-survey on their second contracts, while George Hyde tackled his first. [56] Each man wanted to get the best contracts he could, finish them quickly and make a profit. In the Territory for six months, the surveyors now faced their first Oregon winter.

Chapter 5 – Winter, 1851–Summer, 1852

Chain by Chain

"We had very much wading to do in crossing streams,
sloughs and swamps."

On New Year's Day, 1852, William Ives and his men huddled in tents waiting out the rain. During the short clearing spells of the day that followed, they traced section lines across the sodden ground, but there were more wet days than dry. Rains swelled the Willamette River to its banks, making their work dangerous. Balancing on slick rocks in a fast-moving stream south of Salem, Ives faced near disaster one morning, going "all over under, books, compass and all." Throughout the valley, brimming streams slowed the surveyors' progress and forced them to wade in places where the water was three to four feet deep. The men spent so much time in the water that they cut slits in the instep of their new boots to let the water drain out (Figure 30).[1]

Alder, white ash, and willow grew thick along the streams. Between John Minto's and Virgil Pringle's cultivated fields on Pringle Creek, in what is now urban south Salem, Ives saw an aspen grove, "the first noticed in this country." Nearby, beaver dams backed up small ponds and sloughs along the stream.[2] Ives most likely was unaware that, just a quarter-century earlier, Hudson's Bay Company trappers had nearly exterminated beaver in their relentless harvest of furs in the Oregon Country. Although the mammal's population recovered, its

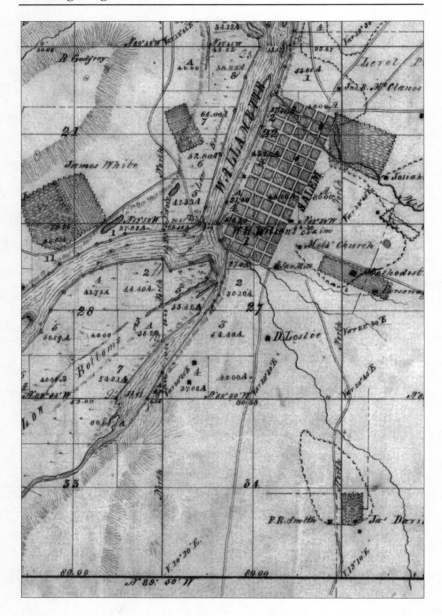

Figure 30. General Land Office Map: Township 7 South, Range 3 West, W. M.;
Salem Oregon and vicinity, Contracts 4 and 9, William Ives, 1852. (USDI Bureau
of Land Management, Portland, Oregon.)

near-annihilation had altered riparian vegetation as streams breached abandoned dams and scoured old meadows, discouraging conifers and promoting instead the stands of willow and alder Ives recorded.

Ives finished his contract on February 11 and returned to Oregon City, carrying to Dr. Evans fossil fragments that the surveyor had found in the Bethel (Eola) Hills northwest of Salem. Evans, who had arrived in Oregon in the fall, planned to spend the spring and summer of 1852 surveying the geology in southwestern Oregon and along the Willamette Meridian from the Columbia River to Puget Sound.[3]

William Ives' training in geology served him well. His mentor, William Burt, had encouraged his surveyors to collect mineral specimens during field work. This extra effort would allow the geologist to "determine the true character of such new districts."[4] His survey finished, Ives sat out the cold days of early March at the Main Street House, tinkering with his compass and reading a book on the history of England. Ordinarily the deputy lost little more than a few days between contracts. Now he sat for weeks while Surveyor General Preston awaited funds and instructions from Washington.

Chilling rains drenched James Freeman north of Albany, where winter flooding turned the low prairies into shallow lakes (Figure 31).[5] In bone-chilling cold, he and his men gave up camping to seek shelter with settlers. The one-room, clay-chinked log houses stood miles apart on land claims 320 to 640 acres in size, and their search for lodging frequently took the men miles away from their survey locations. As Kimball Webster described it, they would stop at one house to hear the farmer say he had "no accommodations to keep folks," and suggest that they try the next place. "The estimate of the distance to the next house," Webster observed, "was almost invariably given at not more than one-half the actual distance."[6] Hiking from farm to farm late at night, the men heard the same response. Some nights no one opened the door. With the small cabins housing several people, the men knew that few families could easily accommodate four tired, hungry men.[7] Webster developed his own technique: "Sometimes I would inquire for a drink of water and gain admission thereby. Once

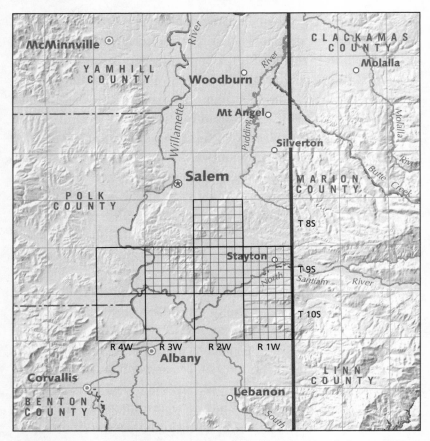

Figure 31. Survey Contract 11(Subdivisions): Township 8 South, Range 2 West; Township 9 South, Ranges 1–3 West; and Township 10 South, Range 1 West, W. M.; James E. Freeman, 1851–1852. (Map image from Stuart Allan, et al., *Atlas of Oregon*. Eugene, Oregon: University of Oregon Press, 2001.)

inside I found the battle more than half won. We could then almost always induce the good people to keep us."[8]

At the beginning of March, Freeman mailed Preston his notes and maps for the completed subdivisions. The weather was still terrible. Unable to cross the Santiam River due to heavy rain and snow, he wrote the surveyor general, "The rivers have been very troublesome as there are few ferries and in high water difficulty rafting them."

Unprepared or undecided settlers made his work even more difficult. The frustrated surveyor complained, "I find it impossible to note many of the claim lines as some of the people . . . have not yet established their lines or if established wish to make changes in some way to claim regular subdivisions, while others are afraid to show their lines, thinking it useless or are afraid they will be obliged to take their land as they have it noted."[9]

Two weeks later, Freeman subdivided Township 9 South, Range 1 West in the vicinity of Stayton, Oregon, noting that "the land lying south of the river [except such areas settled before the treaty] is re-served for the Calapooia [Kalapuya] indians" — referring to provi-sions of the Indian Treaty Act passed by Congress in 1850. By the time the surveyors reached the Willamette Valley, the Kalapuya popu-lation had declined significantly. "When Congress passed the Dona-tion Land Act," historian William Robbins has observed, "the matter of Indian legal claims to land had already been resolved by a previous enactment that authorized the negotiation of treaties with native people in Oregon 'for the Extinguishment of their claims to lands lying west of the Cascade Mountains.'"[10] Freeman counted Thomas Cook, Enoch Huntley, John Whitley, Jackson Neil, and John Montgomery as occu-pants of the reserve prior to the treaty, while settlers claiming land since the treaty were A. Benton, Washington Richison, Nathaniel Huntley, and Thomas Griffith.[11]

As the newcomers cut down trees and turned over the soil, the Kalapuya's world disintegrated. "The Kalapuya's view of their inter-action with the white settlers," historian Peter Boag has noted, "di-rectly concerned their altering relationship with the landscape. The whites 'plough[ed] up the ground,' and thus the Kalapuya's own rela-tionship with it."[12] As Freeman subdivided the land, the few surviving Kalapuya posed no real threat to the settlers. "Their culture . . . in a shambles, their villages destroyed" the starving people begged for food at farmhouses and wandered from place to place.[13] On March 16, Freeman described an encounter that is striking in its inequality: "Yesterday one of the chiefs was inquiring about our doings. I showed him the map & told him by an interpreter that we were making paper talk of it and he said it was all right."[14] In reality, the Kalapuya leader, who had no concept of private land ownership, could not have comprehended the survey's full import. As one settler put it, native peoples were "at a great loss to know how it was that the white men would take compass and chain and go around and cry stick stuck and set up a few stakes and call the land their own."[15]

Meanwhile, east of Salem, George Hyde subdivided along the Pudding and Little Pudding rivers. Sick of the rain, Hyde and his men also boarded with settlers. When an acquaintance visited his camp early in February, Hyde did not invite him to stay since an additional man would make the hunt for shelter even harder. Long days in the wet fields added to the men's discomfort. Lame and unable to work, Hyde wrote Preston after finishing a township north of present-day Silverton, "I have a sore on one of my heels resembling a stone bruise but am told it is produced or caused by being in the water so much." One of his men was ill: "Allen Seymour was attaced with the Cholera Morbus on Sunday Knight last, is now most well and will I think be able to work tomorrow."[16]

⤶

As he tallied the rising number of finished township surveys, Surveyor General Preston triggered the next step in the donation land claim process. On February 10, the *Oregon Spectator* published his announcement that he would now accept settlers' claims notifications. Settlers whose claims had not been measured prior to the public land survey, and those who had not pointed out their claim boundaries to the deputy surveyors, must now have their claims surveyed so that the intersections of the claim lines "with the lines of the *Government survey could be noted* . . . so as to give in their notification a description of the precise tract or tracts of land claimed."[17] The surveyor general's announcement listed the townships for which his office had already approved plats. "All settlers therein," Preston said, "are requested to appear at my office, when and where I shall be prepared to

receive the notifications of their claims respectively." Claimants settling in these townships after the February 10 notice had three months
from their date of settlement to file a notification.[18]

The Donation Land Act gave a settler right to his land from the
date he moved onto it and ploughed his fields. To receive a patent
certificate, a document that granted a marketable title, the settler had
to file an application and an affidavit. Known collectively as the "notification," these papers contained personal information, witnesses'
sworn statements verifying the claimant's date of settlement, citizenship status and continued occupancy, as well as an accurate description of the land. To meet the latter requirement, the settler either had
to measure the land himself or find someone with sufficient surveying skill to mark the claim's boundaries and describe them in writing.[19] Surveyor General Preston did not encourage his deputy surveyors to conduct these private surveys, although political expediency
occasionally led them to do so.

The notifications process confused emigrants and heightened their
anxiety over the land that was their livelihood and future. The system
by which a settler could gain official title was mired in bureaucracy,
while neighbors often passed along information that was distorted by
too many tellings or misinterpreted from the newspapers. The farther
a settler lived from Oregon City or Salem, the harder it was to get
news. When Umpqua Valley residents and new arrivals to still-wild
southern Oregon wrote Preston, they sometimes had to wait months
for a reply.

Hiram Colver, a native of Union County, Ohio, who had settled
his claim early in 1852, wrote Preston "on behalf of Myself and many
other settlers" in the Rogue River Valley. Could the notification be
mailed or did it require appearing in person in Oregon City? Could
one send an agent? Did the surveyor general send patent certificates
through the mail? "As there are no standard lines established by public surveying in this Valley," Colver asked how a settler might describe a survey properly. "Your immediate answer will greatly oblige
Many Settlers in this Valley on whose behalf they are made [and] be

the means of preventing many mistakes, Litigation, and sore Disappointments, in this Rapidly settling country." Colver closed his letter, advising that Preston's prompt reply would "perhaps in the end save the Surveyor General from much that might be unpleasant."[20]

As word of Preston's announcement spread, perplexed claimants northeast of Salem stopped George Hyde in the fields. Elijah Woodward and Asa Simmons asked him to explain the law, and when the surveyor visited George McCorkle's recently finished farmhouse on Howell Prairie, McCorkle asked him for help in surveying his donation land claim. Hyde instead advised Joseph Latshaw, a surveyor who was visiting in camp, to retrieve his compass in Salem and return, whereupon he would give Latshaw all the work he could possibly handle surveying claims. "The people on Howell Prairie," Hyde wrote Preston, "have just come to the conclusion that they cannot give you a description of their claims without a Survey & are all getting them surveyed. [Latshaw] can make from $5 to $8 per day which is much better than paying a board bill in Salem."[21] Latshaw's bonanza exemplifies the way that the rectangular survey's rapid expansion created employment opportunities for men trained in surveying. Settlers needed their land measured to make their notifications to the surveyor general's office, all in advance of the final formal donation land claim surveys. There weren't enough skilled engineers to go around, and since the government deputies did not generally undertake private contracts, a man with adequate surveying skills could count on finding work.

Spring, 1852

After his winter hiatus, Butler Ives left Oregon City on February 17, 1852, to resume surveying in the Molalla country. With George McFall and three other men, he headed twenty miles south to the vicinity of what is now Marquam. After ten days of intermittent sun and clouds, the weather turned bad. On Sunday, March 1, a day of intermittent snowstorms, Ives went deer hunting in the afternoon, shooting at several and wounding one.[22] The surveyors hunted for sport and to supplement their food supply. In the early years of settlement, both black-tailed deer and white-tailed deer roamed along the western edge of the Cascade Range from Puget Sound to southern Oregon. Both species thrived in the eastern reaches of the Willamette Valley and, although Ives, who wore glasses, missed most of his shots on this expedition, the surveyors were usually able to bring down deer.

Two days later, in snow that covered the ground, Ives' again fell ill with symptoms of the malaria he had suffered months earlier. Shivering from chills, on March 4 he dragged four and one-half miles to finish surveying Township 5 South. Making no mention of his illness, the next day he wrote Surveyor General Preston from camp near Rock Creek west of Molalla, Oregon, that he would work as fast the weather permitted.[23] Ives sent William Potter to Oregon City with the letter and field notes for Preston. That night Ives and the rest of the company moved camp into Township 4 South, where settler Andrew Gribble and Nancy, his seventeen-year-old bride, took the ailing surveyor into their home. Gribble, born in 1822 in North Carolina, had settled his 602-acre land claim at what is now Macksburg, southeast of Canby, Oregon, in November 1850.[24]

Still suffering chills and fever, Ives was very weak when the company stopped to camp near the Molalla River on March 22. After one long day in the rain, the men tried to ford the river, only to find that they had been stranded from their camp by the fast-rising water. With no farmhouses nearby, they slept out without blankets or food. Hurrying to finish the survey, the company returned to Oregon City, where Ives joined his brother at the house they rented from Mrs. Thurston and took a few days to recuperate from his illness.[25]

Sufficiently recovered to resume work, Ives returned to the field on March 27 and, over the next month, subdivided Township 1 South, Range 1 East, a heavily settled township that encompassed the city of Milwaukie and part of Portland.[26] The surveyor found the township booming. "A large amount of improvements are being made both in the towns & farming community with a spirit that bodes fair to remove the natural obstructions to agriculture & its growing commerce."[27]

Over the next eight weeks, Ives endured incessant rain to subdivide two townships south of the present-day Lake Oswego and managed to complete his survey in the second week of June. He found that only a few of the settlers in the area had paid to have their claim boundaries "deffinitely settled at time of survey," saving money instead by relying on the official section surveys to locate their claim lines.[28]

In Oregon City, Preston tapped Robert Elder to join the five surveyors already in the field. Preston knew Elder from association with him on the Illinois-Michigan Canal and, as noted earlier, had invited the civil engineer to come to Oregon. Elder resigned as chief clerk in the surveyor general's office and took a survey for ten townships extending south of Hillsboro and Forest Grove to the vicinity of Amity, Oregon. With Kimball Webster as his compassman, Elder started out on April 30, 1852, with a company that included Henry S. Gile of Maine and Matthew O. C. Murphy as chainmen, and Andrew Murphy, Matthew's younger brother, and James M. Fudge as axemen.[29] Before leaving Oregon City, Robert Elder arranged to team with Webster on a later contract (Number 20) to subdivide these same townships. This joint arrangement allowed the surveyors to accept contracts of ten townships instead of the five usually awarded in a single subdivision contract.[30] Elder and Webster were the first deputies in Oregon to officially share a survey, although Butler Ives had assisted his older brother and Webster had previously worked closely with James Freeman. William Ives worked solo contracts and, with two exceptions, so did Butler Ives. George Hyde teamed on two occasions with another deputy, but most of the time, the surveyors took the full work load, as well as all the risk and profit, of individual contracts.

In the late spring of 1852, James Freeman hurried on his third post-meridian contract, subdividing five more of the townships for which he had surveyed boundaries in an area that encompassed the Forks of the Santiam River and the Willamette River from Albany to Salem.[31] On the Santiam River in Township 10 South 3 West, he visited two small settlements about two miles downstream from present-day Jefferson. Today, automobiles on the Interstate 5 bridge rush over the river where Santiam City and Syracuse, both destroyed in the great flood of 1861, now lie buried (Figure 32). Freeman's description fleshes out our knowledge of these two struggling, small communities long-disappeared from the landscape. Santiam City sprang up on the north bank of the river on Samuel S. Miller's donation land claim. Milton Hale built a ferry on the Santiam in 1846 and platted Syracuse on the south bank opposite Miller's land. "Santiam City on the right bank," Freeman observed, "contains two stores, 3 dwelling houses, one saw-mill & three shops. Syracuse on the opposite or left bank contains 2 houses. Neither of these cities," he ventured, "are in a very flourish-ing condition."[32] By the time of Freeman's visit, Santiam City, the more aggressively developed of the two towns, had claimed the post office and eclipsed smaller Syracuse.

Crossing open prairie along the Willamette River south of present-day Independence in early July, Freeman and his company visited another struggling settlement promoted to serve travelers and settlers. "At Hall's Ferry on left [east] bank of Willamette River is a ware-house & good steamboat landing. A few town lots have been sur-veyed called Buena Vista."[33] Reason B. Hall had opened a ferry on the river two years before and James A. O'Neal had built the mercan-tile structure. Hall's ferry was one of many that were plying Territo-rial rivers at the time (Figure 33). Indeed, as *Chaining Oregon* goes to press in the spring of 2008, the Buena Vista Ferry, operated by Marion County from May through October, still carries travelers across the Willamette River.

At the ferry landings, licensed operators made a good business transporting foot passengers, wagons, horses, mules, cattle, sheep, and freight across the streams. Ferry transportation could be dangerous. High water made the crafts unmanageable, ropes broke, and shifting loads plunged passengers and animals into the water. In spite of the risks and the cost, ferries such as White's at Salem, Griswold's on the Sandy, Skinner's on the West Fork Willamette, and Knott's on the South Umpqua River were essential to travel in the Oregon Country. The deputy surveyors used them frequently, paying to transport crews

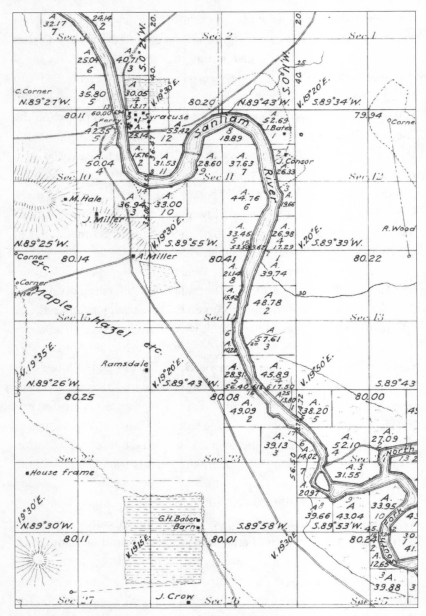

Figure 32. General Land Office Map: Township 10 South, Range 3 West, W. M.; Santiam River and Syracuse, Contracts 5 and 16, James E. Freeman, 1852. (USDI Bureau of Land Management, Portland, Oregon.)

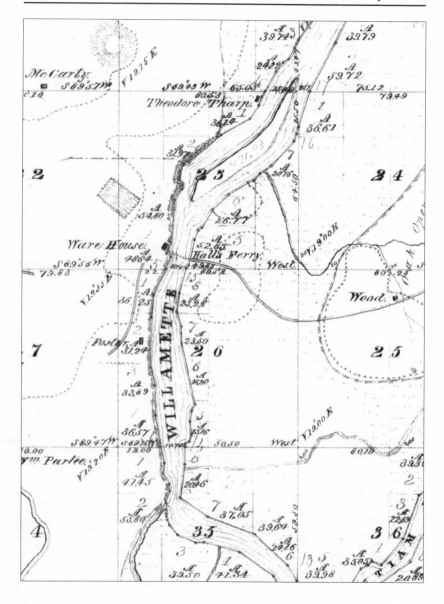

Figure 33. General Land Office Map: Township 9 South, Range 4 West, W. M.;
Willamette River, Buena Vista and Halls Ferry, Contracts 5 and 16, James E.
Freeman, 1852. (USDI Bureau of Land Management, Portland, Oregon.)

107

and pack animals to new survey locations or to move between townships on their contracts.

While James Freeman worked southwest of Salem, William Ives took Contract 12, his third since finishing the meridian and base line (Figure 34). On April 10, his thirty-fifth birthday, he left Oregon City to survey ten townships in a twelve-mile-wide swath extending from the base line across the Tualatin and Willamette rivers to the south boundary of Township 5 in the vicinity of Mount Angel, Oregon. Ives

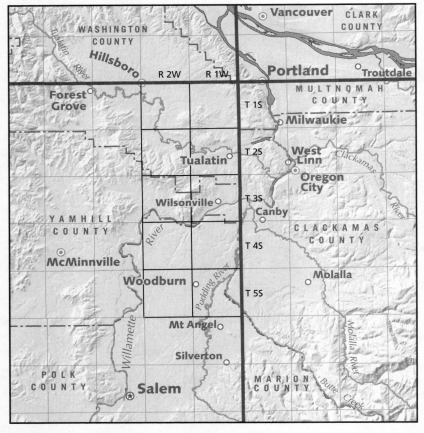

Figure 34. Survey Contract 12: Townships 1–5 South, Ranges 1–2 West, W. M.; Salem, Oregon, vicinity, William Ives, 1852. (Map image from Stuart Allan, et al., *Atlas of Oregon.* Eugene, Oregon: University of Oregon Press, 2001.)

and his company of Samuel Gatton, Isaiah W. Case, James D. Price, William Gilcrist, Sylvester Cannon, and John A. Harry started work near Hillsboro. Cannon and Case were new to Ives' company as was Harry, an Indiana native who had arrived in Oregon in October, 1851.[34]

Ives' new survey encompassed French Prairie, an expanse of fertile agricultural land between the Willamette and Pudding rivers and one of the longest-settled areas of the Oregon Country (Figure 35). In this region, French Canadian settlers, many of them former HBC employees who had taken most of the existing land claims prior to the public land survey, described their boundaries under the metes-and-bounds system using landmarks such as trees, rocks, and streams. These French Canadian settlers carried with them to Oregon the tradition of shaping their land claims as "long lots," or deep narrow properties, with the long dimensions situated perpendicular to a river or stream which provided convenient access to transportation and commerce (Figure 36).[35] In the early 1840s, André Longtain and Robert Newell had registered claims on French Prairie under the laws of the Provisional Government; Newell represented his district in the Provisional Government legislature each year from 1844 to 1848 and served twice as speaker of the Assembly (Figure 37).[36] The town of Champoeg, seat of pre-territorial provisional government, and now an Oregon heritage area and state park, developed on Newell's and Longtain's adjacent land along the broad plain flanking the south bank of the Willamette. There, Ives counted twelve log buildings on the prairie extending between the Hudson's Bay Company warehouse on the west to Robert Newell's improvements about one-half mile to the east.[37] Built in 1841–1843, the warehouse stored the HBC's Willamette Valley wheat for later sale (Figure 38). In the western reach of Ives' contract area was the settlement of St. Paul, where the 1839 Roman Catholic Mission and Oregon's first Roman Catholic Church (St. Paul's, 1846) served the resident HBC families.[38]

On their best days the company made six miles, and on their slowest they covered four. Ives finished the boundary survey for Township 4 South, Range 2 West in mid-May and, crossing the Willamette

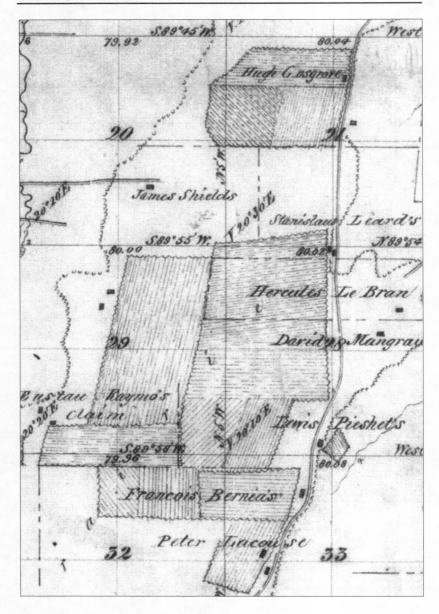

Figure 35. General Land Office Map: Township 4 South, Range 2 West, W. M.; French Prairie, Contracts 12 and 18, William Ives, 1852. (USDI Bureau of Land Management, Portland, Oregon.)

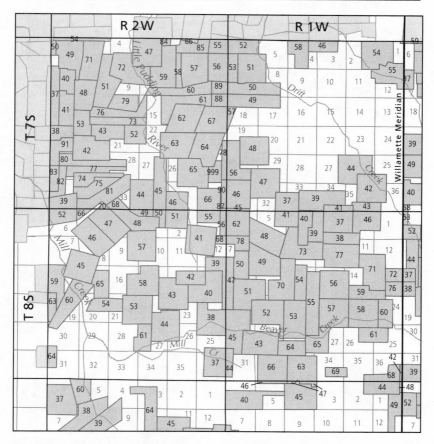

Figure 36. Detail of claims settled prior to the Donation Land Claim Act of 1850. (Map image from Stuart Allan, et al., *Atlas of Oregon.* Eugene, Oregon: University of Oregon Press, 2001.)

on the ferry at Champoeg, returned briefly to Oregon City. He turned in his notes to the surveyor general's office on May 19 and that evening bought four drinks of liquor to celebrate.[39] Two days later, on May 21, Preston accepted Ives' notes and maps for Contract 12 and paid him $1,840.11 for his work. The surveyor general immediately awarded the surveyor a new contract to subdivide some of the townships for which he had just finished boundary lines. After spending $322 for

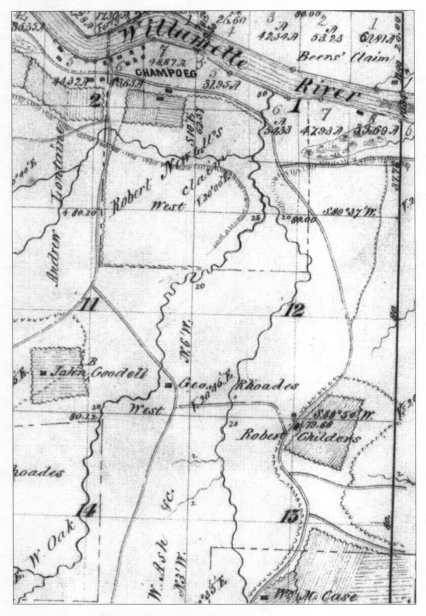

Figure 37. General Land Office Map: Township 4 South, Range 2 West, W. M.; Champoeg; Robert Newell's land claim, Contracts 12 and 18, William Ives, 1852. (USDI Bureau of Land Management, Portland, Oregon.)

Figure 38. Champoeg Village, 1851. (George Gibbs sketch, Oregon Historical Society, OrHi 44495-A.)

supplies; $600 for wages, and $3.50 to the Main Street House, Ives realized a profit of $914.21 on Contract 12. He left Oregon City at 3 P.M. to return to his company, arriving at Champoeg just before 10 P.M.[40]

Now, the surveyor turned to subdividing, camping most nights near land claims where the company supplemented their staple provisions with butter, peas, milk, fresh beef, and eggs purchased from nearby farmers. Other nights, the men set up camp near town where they could buy flour, sugar, dried fruit, and tea. The men camped six nights at Champoeg, three nights near Indiana native William M. Case's farm — Case had settled in Oregon in 1844 — and two nights near Alphonso Rhoades' claim, where the New York native had settled in 1850.[41] Ives put William Gilcrist in charge of moving camp until June 19 when he gave James Price the job. The campman's role was an important one requiring expertise with both a compass and pack animals. Gilcrist, whom Ives would soon fire for being "careless & lazy," temporarily worked on the survey line.[42]

On June 5, William Ives broke from his tight schedule to survey Robert Newell's one-mile-square land claim, one of the few recorded instances in which deputy surveyors surveyed private claims. By surveying Newell's claim, Ives accommodated the prominent settler who signed the surveyor's security bonds for Contract 18. The company worked on Sunday to finish the claim survey, foregoing their usual resting Sabbath. After six days of hard work, both William and Butler Ives usually gave their men Sundays off and spent time resting, writing letters, or reading. The brothers always carried a volume of Shakespeare or a history book with them. They also attended services at Methodist, Presbyterian, or Episcopal churches.[43] On one occasion early in June, William Ives and four of his men left camp near William Elliott's claim east of Butte Creek, southeast of present-day Hubbard, Oregon, and walked six miles to attend a camp meeting at Samuel Allen's on Abiqua Creek. On Sunday afternoon, after the meeting they walked back to camp near Elliot's farm.[44]

When they had finished surveying Newell's claim, Ives and his men took an extra day off to vote. On Monday, June 7, 1852, citizens throughout the territory went to the polls to elect representatives to the legislature and to local offices. Ives and his men cast their ballots; this was chainman Gatton's first election. Later in the day, the deputy surveyor and his men celebrated by sharing a quart of whiskey to mark the occasion. "Staid at Champoeg wrote notes and staid at the election to see a few drunken French and Irish," Ives wrote in his journal that evening.[45]

After another full week of work, Ives sent John Harry to Oregon City with the field notes and sketch maps for the French Prairie country. Harry returned the next day with several letters and a stack of newspapers.[46] One of William Ives letters was from Loren Williams, who had stayed behind in Acapulco the previous year to recover from illness while the rest of the company went on to San Francisco. Apparently surrendering to uncontrollable temptation, Robert Hunt, whom Ives had left to care for the sick man, had deserted his charge and used the money Ives left for the patient's care on his own needs and pleasures.

"Your old *friend & fellow traveler*, Bob Hunt," Butler Ives would write his brother years later, has gone through "the general mill of old California . . . gambling, drinking mining," and claimed that "he was taken sick himself at Acapulco & was obliged to payout the money for *medical* services."[47] Williams, who now lived on the Umpqua River, had accompanied William G. T'Vault's exploring expedition the previous September and narrowly escaped death when Indians murdered most of the party. Williams wrote Ives the details of his ordeal and asked the surveyor for help. "I would call on you once more for a little help again if you were where you could . . . write to me as soon as you get this for if I live I shall be a here a long while yet."[48]

Summer, 1852

Ives' company, down to three men since Gilcrist's departure and John Harry's sudden illness forced him to stop work, took July 3 off in Champoeg to celebrate Independence Day, a holiday enjoyed exuberantly throughout the Territory. The crew joined in the festivities that included a dinner, rounds of public toasts, and an evening ball.[49] The survey party had returned to work and was camped on the south bank of the Willamette River a few miles east of Champoeg when N. DuBois and George H. Belden, clerks in the surveyor general's office who had business in Champoeg, stopped to visit. William Ives greeted his callers cordially, for he and the other surveyors maintained good working relationships with the clerks on whom they depended for fair, accurate, and timely review of their field notes.[50] Recent additions to Preston's staff, DuBois had arrived in the northwest in 1851 and worked as a surveyor and draftsman in Milwaukie before joining the office in the spring of 1852, while George Belden, a skilled engineer, joined the office early in the year and occasionally did some surveying in the field.[51] The two new clerks joined Wells Lake on Preston's staff. Born March 8, 1816, in Mount Morris, New York, he moved to Chicago as a young man and, by 1850, promoted transportation of wheat by railroad in that region. In California in late 1851, Lake accepted Surveyor General Preston's offer of work in Oregon City.[52]

William Ives finished Contract 18 on August 12 and returned to Oregon City. Preston paid him $3,194.93 and Ives reimbursed his crew, paying only for the days they actually worked, but covering any travel and lodging expenses incurred on the contract. Ives gave Samuel Gatton a little extra. Gatton, who had taken on extra duties after Gilcrist's departure, earned $65.00 per month for three months and three days, and $80.00 per month for one month and nine days. After tallying $298.83 for supplies and expenses and $910.56 for labor, Ives realized a profit of $1,985.54.[53]

As his brother finished Contract 18, Butler Ives hurried to finish his own subdivision survey in the Tualatin River Valley.[54] With George McFall, William Potter, and three others, he worked south and west of Portland, hindered tremendously by poor weather and inadequate axe men, to finish his two townships in mid-August. Ives agreed to take a special contract, and he and McFall, with one of Joseph Hunt's men, surveyed the boundaries of the US Ordnance Depot Reserve located in the vicinity of present-day Oaks Bottom Wildlife Refuge in southeast Portland. Anticipating future need, the US government had withdrawn the reserve from the public domain for military use, making the land no longer open to settlement.

When he got back to Oregon City on August 26, Butler Ives found his brother "fitting up to leave for the states."[55] The younger man had known since spring that William Ives would not spend another winter in the West. The surveyor general had issued nineteen township boundary or subdivision contracts since August 1851 — William Ives had reaped four of them — and many more surveys were needed in agricultural areas north of the Columbia River, in the Rogue River Valley, and in the small valleys and foothills west of the Willamette Valley. William Ives' departure would leave just six US deputy surveyors working in the Territory. To Butler Ives, the future in Oregon without his brother appeared to be bleak and lonely.

Chapter 6 – Fall, 1852

Politics and Patronage

"I almost wore my legs out in Oregon."

William Ives was tired. "I almost wore my legs out in Oregon," he would say in later years, and he deeply resented the General Land Office's tardy payments on his contracts. Preston paid Ives $500 at the end of August but still owed him over $5,000.[1] But more compelling than money or sore joints, Ives was going home to marry Sarah Hyde. He paid Mrs. Thurston $30 for his part of a year's rent at the house in Linn City. To keep up with the Territory's stimulating political scene, the surveyor paid to have six months of the *Oregon Statesman* mailed to Detroit and canceled his subscription to the *Oregonian*. William Ives sold Butler Ives his English nautical almanac, a copy of Nathaniel Bowditch's *New American Practical Navigator*, and his French dictionary, and gave his younger brother power of attorney to transact his unsettled business with the surveyor general's office.[2]

Reluctant to end his once-in-a-lifetime sojourn without seeing the gold country, Ives chose the land route to San Francisco. Since his survey contracts had extended only as far south as Albany, Oregon, this itinerary would allow him to see more of the West. On the morning of August 31, William and Butler Ives bid each other goodbye. The parting cannot have been easy, for they had been together — the older brother a mentor to the younger — for six years. William Ives' reflection turned to exasperation, however, when, four miles outside Oregon City, his mule ran off. His tracking efforts fruitless, Ives returned to Oregon City, borrowed his brother's horse and went back to search for the wandering animal. He bought a horse to replace the mule and, as he was leading Butler's horse back toward Oregon City, he encountered the missing animal minus the baggage. Ives started out again the next day toward California, this time making forty miles to reach Salem for the night.[3]

The day William Ives left Oregon City for good, Preston awarded Butler Ives Contract 22 to survey boundaries for ten townships at $17 per mile "on conditions that [he] not call for the pay till appropriations had been made by Congress and the money received at the surveyor general's office." This restriction on payment emphasized Preston's struggle to keep the surveys going despite tardy appropriations from Washington. For his new contract in the Willamette Valley from the West Tualatin Plains to a few miles north of Marysville, Ives bought a bay horse and stocked up on provisions of bacon, pork, cornmeal, beans, and sugar. Rounding out his outfit, Ives purchased a Gunter's chain, new buckskin trousers, cowhide boots, candles, and tin kettles.[4] He rehired Keen, McFall, and Potter, and added Orlando Neal, F. O. Collins, and Alston Martindale to his crew. Then, just before his departure, the court summoned Ives for jury duty on a murder trial. Like many responsible but over-committed citizens, Ives appealed the appointment and to his great relief the court excused him. Leading pack mules, the company covered forty-six miles by way of Hillsboro and Forest Grove in two days. On September 9, Ives sent McFall ahead with the horses to establish the night's camp near present-day Gaston. The next morning, the men loaded a

Figure 39 (at right). Survey Contract 22: Townships 1–10 South, Range 5 West, W. M.; Butler Ives, 1852. (Map image from Stuart Allan, et al., *Atlas of Oregon.* Eugene, Oregon: University of Oregon Press, 2001.)

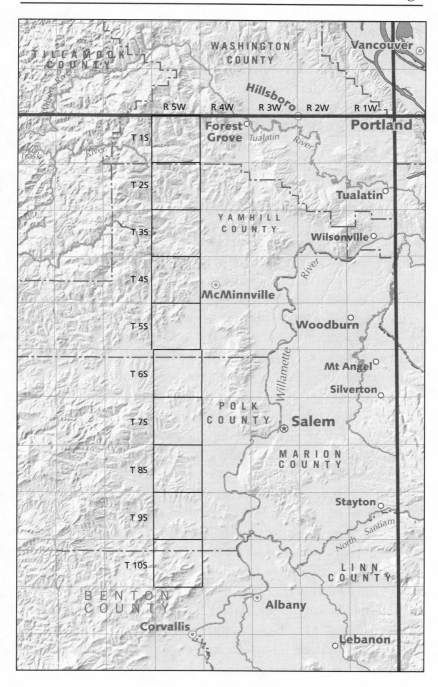

week's provisions on their backs and headed into the Coast Range (Figure 39).

Two days later, McFall fell ill with chills and fever, and Ives left him to rest in Collins' care.[5] Now down two men, the company cut through tangled thickets, working their way south. On September 26, they reached Richard Combs' claim west of McMinnville in Yamhill County. Kentuckian Combs, who had arrived in Oregon in 1847 and soon settled his claim, supplied the survey party with milk, butter, potatoes, and tomatoes. The next day, the men surveyed into the southeast corner of the township and camped near Henry Warren's house. Warren, a native of Nova Scotia, and his wife, Nancy, also had arrived in Oregon in 1847 and farmed a 640-acre land claim. New Englander Butler Ives thought the Warrens "fine folks & Yankees" whose eastern accents stood out in the Territory where many of the settlers hailed from the Border States.[6] Ives and his men camped near several other settlers, including Peleg Hibbard, also a native of Canada, and Missourians Rufus and Evaline Riggs. The company finished a week later and, on October 14, Ives sent McFall and Neal across the Willamette River to camp until the start of the next contract, while he and Potter started for Oregon City on horseback. That night, the deputy and his assistant broke up their trip with a night's lodging at a hotel on French Prairie where, Ives reported, "the French and their squaws had a drunken dance . . . in the evening full of fun frolich & fighting."[7]

Ives reached Oregon City the next day, recopied his field notes and sketch maps and turned them in to the surveyor general's office. Although he was anxious to return to work, the surveyor lagged behind, awaiting a ship carrying the US mail from San Francisco. It arrived almost a week late, on October 25. Ives picked up two copies of the Detroit *Weekly Free Press* and the September issue of the *Eclectic Magazine of Foreign Literature, Science and Art*. The latter edition, a thick, leather-bound volume, reprinted articles and fine engraving from the best English periodicals on astronomy, biography, geology, the arts, archaeology, and history. Best of all, the postmaster handed him a letter from William Ives.[8]

∽

The trip up the Willamette Valley took William Ives six days. On September 4, he passed through Marysville and the next day reached the north slope of the Calapooia Mountains. On September 7, he made twenty-five miles along the territorial road to cross the north branch of the Umpqua River by boat. The following day, Ives went thirty miles — to near present-day Canyonville — and stayed the night with a settler before entering "the Canyon" described by the surveyor as "a rocky ravine with rocky mountains on each side . . . [where] it is difficult for wagons to pass & bad for horses or mules with packs." He stayed September 9 on Grave Creek at Bates and Twogood's inn near the main road in the narrow Grave Creek Valley that lay between two rugged mountain ridges, present-day Smith Hill Summit and Sexton Mountain Pass. James Twogood had constructed the log house six months earlier and, with A. S. Bates, offered food and lodging to travelers while he also earned a living by hauling supplies to the mines. Families, packers, and miners crowded around the inn where Ives counted "5 wagons with families camped with Horses and cattle bound for Rogue River gold mines. Thirty-four mules with 7 men camped same place going north."[9]

William Ives reached the Rogue River on September 11, in sunshine and smoke. Ferrying across the fast-moving stream, he reached Jacksonville in the Bear Creek Valley, which comprises the main part of the Rogue River Valley, at 5 o'clock that evening and rented a room at Evans' boarding house. Looking around, Ives saw a rough little town with:

> about 30 small one story houses built of logs and split boards besides . . . some tents. Numbers of the Houses are boarding houses 2 or 3 of them are respectable about 10 of them keep a bar One a ball alley and gambling tables. There is also many other gambling tables in the middle of the village and a Whorehouse with three women where they have fiddling dancing drinking swearing and fighting. Most of the shops and houses are kept open Sunday.[10]

Figure 40. View of Jacksonville, Oregon, 1854. (Southern Oregon Historical Society Photograph #805.)

In town on a Saturday night, Ives bought himself two drinks. The entertainment grew livelier with the arrival that evening of "three Calafornia Ladies of accommodation . . . who created considerable excitement among the miners. There was fiddling dancing drinking etc. where they stopped" (Figure 40).[11] Staying in town over Sunday, September 12, Ives set out Monday to explore the mines west of Jacksonville, "[I] went up the west gulch or ravine where they were mining for gold All of the way for 2 or 3 miles. Digging it from the gravel stones & boulders on the surface of the rock called the bedrock where most of the gold is to be found." Ives got his own gold in town. "I bought one ounce of gold for $17 for specimens I also took another ounce for $16 to make change."[12] That afternoon he left Jacksonville, riding fifteen miles south to camp near a "new sawmill" at what is present-day Ashland. The next day Ives crossed the Siskiyou Mountains into California and was gone from Oregon Territory.[13]

〜

In late August and early September, Surveyor General Preston issued the first of three contracts for the official survey of individual donation land claims, an important step in finalizing the surveys. When settlers had filed notifications for all the claims in a township, Preston mapped their locations. Next, the surveyor general awarded a "claim contract" for surveying the approved claims in that township and, after resolving any last boundary conflicts, issued each claimant a certificate. The document gave the settler a right to a patent and in most instances sufficed as evidence of legal ownership until the General Land Office issued the final patent.[14] The claims contracts called for the appointment of deputy surveyors especially for the task; the General Land Office paid them standard rates per mile for their work, forbidding any double fee for surveying the same line twice where adjacent claims had a common boundary.

Joseph Latshaw, Jerome Green, and Israel Mitchell got the first three contracts, all in the Willamette Valley. "The contracts for the Donation Land Claim Surveys," historian C. Albert White has noted, "were usually given to Deputy Surveyors who lived in the local area. Since these were not lucrative contracts, it was essential to keep costs at a minimum. The field crews were often local residents and/or the claimants themselves assisting the Deputy."[15] Joseph Latshaw had surveyed claims for notifications — George Hyde had set him up on Howell Prairie — and Israel Mitchell, recently elected to the Lower House from Washington County, had assisted James Freeman as chainman on the final segment of the meridian survey. Among his assigned six townships, Latshaw surveyed twenty claims in the vicinity of Salem (Township 7 South, Range 3 West) between September 1 and October 9, 1852. He worked in the field for half the time, surveying one claim on most days and occasionally two. On other days, he copied his field notes for clusters of claims completed, refined his sketch maps, and turned them in before going out again.[16] Jerome B. Green surveyed six townships between Salem and the foothills of the Cascades, while Israel Mitchell surveyed Tualatin River Valley claims southwest of Portland.[17]

The Donation Land Act required that settlers pay for these official surveys of their claims.[18] Claimants who had already paid once to have their land measured for the notifications complained bitterly about the additional charges. At the same time, some land claimants slowed the entire process by neglecting to submit their notifications. As historian Stephen Dow Beckham has observed, "Once settlers had filed their notices of entry for either donation land claims or 'preemption' claims, they were extremely slow to take title. The pioneer generation was reluctant to procure a record of ownership for their lands because once the land was patented, it was subject to taxation."[19] As long as a settler's land was registered as a claim with the surveyor general, or later with the land office, he held a protected right to the property. Not until he wished to divide his land for sale did the claimant require a patent.

As fall approached, new settlers poured in to the Willamette Valley. "Portland, Milwaukie & Oregon City are full of emigrants in fact the whole valley is now pretty well filled up," Butler Ives observed (Figure 41).[20] An estimated 10,000 settlers arrived in the season of 1852, nearly triple the number of 3,600 counted for 1851. The resulting responsibilities for managing township surveys, donation land claim surveys, and correspondence overwhelmed the surveyor general's office. Because the Donation Land Act had not provided for land offices, registrars, and receivers, Preston had to do it all, acting "as a land office, Surveyor General, and as a judge to adjudicate any conflicting claim disputes."[21] He supervised the checking, copying, and recording of all township, subdivision, and donation land claim surveys; issued contracts to deputy surveyors; and reported to the Commissioner of the General Land Office.

As he had since arriving in Oregon City, Preston read the stacks of letters that poured in from settlers like James L. Cooper near present-day Monmouth, southwest of Salem. "The law gives to a man and his wife each claimes," Cooper began. "[I]f the wife dies and the man marries is his second wife intitled to her claim . . . and if a married man with children . . . has taken a claim and is at work on it bilding to

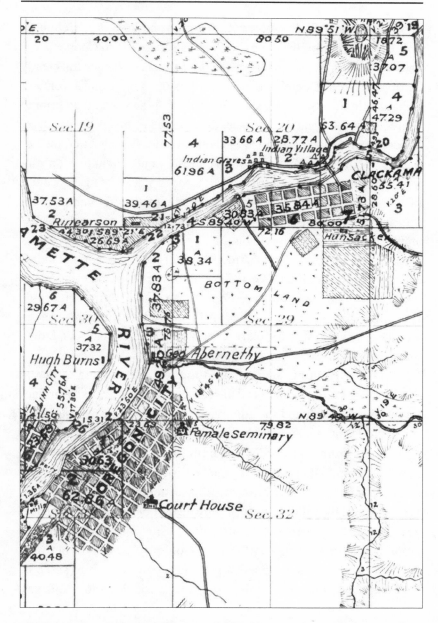

Figure 41. General Land Office Map: Township 2 South, Range 2 East, W. M.; Oregon City, Clackamas City, Oregon, 1852. (USDI Bureau of Land Management, Portland, Oregon.)

go on and his wife dies before moving on the land is her children intitled to her part of the claim."[22] Preston's response to Cooper is not known to have survived, but in responding, the surveyor general clearly had to address the Donation Land Act provisions concerning a woman's right to hold title. Section 8 specified only that if a settler died prior to his required four years' residence, that "all the rights of the deceased under this act, shall descend to the heirs at law of such settler, including the widow." The text did not give Preston much help in determining the legal rights of children of Mr. Cooper's dead spouse, or those of his second (living) wife. Other sections of the Act came up for scrutiny in assessing various personal situations. Section 4, for example, defined those eligible for grants of land as "every white settler or occupant of public lands, American half-breed Indians included, above the age of eighteen years." Many of the older male settlers in the French Prairie region were French Canadians, former Hudson's Bay employees who had married or long lived with Indian women, and they wrote to ask for clarification of the legal status of Indian wives in Oregon.

In October, in his annual report to the General Land Office in Washington, Preston announced the completed survey of exterior lines for sixty townships and the subdivision of fifty-five of those townships. Notifications had come in from 777 settlers holding 640-acre claims, and from 202 residents claiming 320 acres under the Act's fourth section. Eight settlers had filed notifications for 320 acres, and twenty claimed 160 acres under the fifth section, for a total of 567,600 acres claimed by 1,007 settlers (Figure 42).[23] For the upcoming fiscal year, the surveyor general recommended using one-third of the appropriations for surveys "north of the Columbia, one-third between the Columbia and the Calapooya mountains, and one-third south of the Calapooya mountains." His cost estimate asked for $6,500 to pay

Figure 42 (at right). Diagram of a portion of Oregon Territory, 1852, published by John B. Preston, Oregon Surveyor General. (Map image from Stuart Allan, et al., *Atlas of Oregon*. Eugene, Oregon: University of Oregon Press, 2001.)

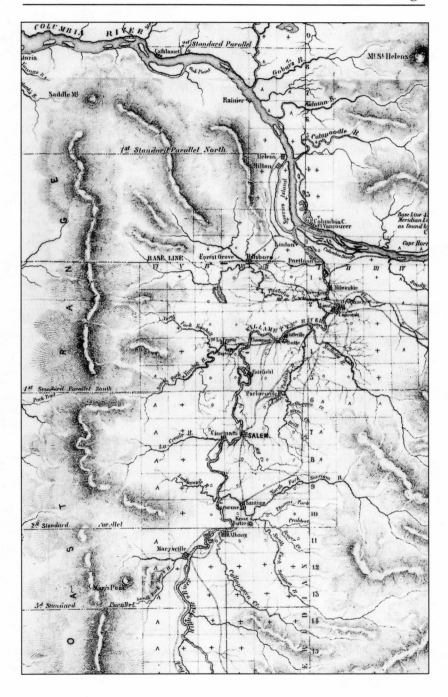

himself and clerks in the office and $8,100 for additional clerks to make up for inadequate regular appropriations. According to Preston, the cost of surveying "standard parallel, township, and subdivision lines equal to eighty townships, say 6,400 miles, at a rate not exceeding $12 per mile" came to $76,800. He estimated the expense for extending the meridian line north into what is now Washington state and south into the Rogue country at a rate not to exceed $25 per mile at $4,000, and added $9,250 for office rent, fuel, books, and supplies.[24] On November 20, the *Columbian* in Olympia published Preston's announcement that surveyors would begin work north of the river during 1853. According to the newspaper, the surveyor general slated for survey four townships north of Vancouver, eight townships north of the Cowlitz River, as well as others farther north.[25]

On December 1, 1852, Preston awarded five survey contracts, the last of eighteen issued during the calendar year — one each to James Freeman, Butler Ives, Joseph Hunt, Robert Elder, and Kimball Webster.[26] When Webster approached the surveyor general for a contract of his own in November, 1852, he encountered firsthand the political realities faced by surveyors in Oregon. Preston asked Webster about his political affiliation. Answering that he was a Democrat, Webster asked if the affiliation would hurt his chances of getting a contract. Preston explained that while it would not prejudice him, other surveyors general tended to award contracts only "to deputies of their own political faith." Most of his surveyors, Whig Preston told Webster, were Democrats, "Mr. Freeman, William Ives, and his brother, Butler Ives, and Joseph Hunt were all of that party."[27] True to his word, Preston awarded Webster Contract 27 to subdivide townships in the upper Calapooia River Valley in the vicinity of present-day Brownsville and in the western Willamette Valley.[28]

⌒

Butler Ives, who had won his previous contract only by promising not to seek payment until moneys were available, wrote William Ives on December 2, "A double mail has just arrived after a delay of nearly a month; bringing to the Surveyor General money & instructions; for which we have been waiting for the last two months."[29] Desperate for money from Washington, Preston had written Commissioner Butterfield, "On the 23rd of June I made a request for the balance of the appropriation of 3 Mar. 1851 — I am not in receipt of any money since that date and fear there has been some miscarriage of the money by the mails. We are all in *want*, Deputy Surveyors, clerks in the office and myself."[30] Although he was relieved to finally have a contract, Ives' letter reveals his growing resentment of Preston's partiality toward some of the surveyors. George Hyde's familial relationship gave him a personal advantage. James Freeman, who had grown close to Preston on the trip west, now enjoyed his favor. That Freeman's and Preston's relationship was closer than those of Preston and his other deputies is further suggested by Freeman's signature on letters to Preston as "your friend" while the other surveyors signed more formally as "respectfully yours" and "your obedient servant." Freeman had finished Contract 24 in early November, 1852, and returned to Oregon City where Preston allowed him to begin his next survey, despite the lack of funding. On December 1, when the surveyor general formally issued the new contracts, Freeman was already at work.[31] "Freeman has been in about three weeks & has yet about forty miles of his township lines to run," Ives told his brother, "He is going out a subdividing & leave the town lines till good weather. Hyde finished about a month after I did, & is now subdividing on his own hook or without the form of a contract . . . Hyde and Freeman have commenced while the rest of us have to wait the motion of mails."[32]

Oregon's typical winter weather had not changed, Butler Ives reported. "The rainy season set in this fall about the 20th of October & since that time it has been a perfect drizzle scarcely sun enough to adjust a compass & the streams are all booming with water & there is mud in plenty." Concerning his brother's long overdue pay, Ives wrote, "Inclosed I send you a draft of three thousand dollars payable to your order; drawn by Jno B. Preston on Messrs. Bradman & Co., New York. Preston wanted fifteen days time allowed so as to be sure &

have the necessary funds deposited . . . The remainder of the money due you at the office will not be paid till returns are received from their accounts of the last quarter; which may be two months yet." Of his own survey contract, Butler Ives said, "I am still further up in townships from 15 to 18 South inclusive of range 4 West & one of five west Willamette river in two townships All wet work, but *I guess I can do it.*"[33]

ᔄ

At the end of the year, Preston dodged the verbal assaults of his political enemies on national and regional fronts. As virulent animosity flared between Oregon Democrats and Whigs, *Oregon Statesman* editor Bush attacked Governor Gaines in the press, while Thomas J. Dryer, editor of the Portland *Oregonian,* supported the governor and his associates. Preston's close relationship with the Whig governor — Gaines married Margaret Wands in the Preston's home in late November — made the surveyor general even more vulnerable in the territory's highly charged political climate. [34]

In mid-December, the bark *Charles Devens* arrived from San Francisco with news of Franklin Pierce's election on November 4, 1852, an event that would change the national administration from Whig to Democrat.[35] Firmly linked with concepts of friendship and loyalty in nineteenth-century America, political patronage guaranteed the ousting of Fillmore's appointees to make way for men of the new president's choosing. From the week before Christmas, when he heard the election results, John Preston expected to lose his job.

Chapter 7 – Winter and Spring, 1853

Surveyor General

"He advised his friends to 'make hay when the sun shines.'"

"Oregon is still Oregon," Butler Ives wrote William Ives early in the year, "though the month of December it seemed more like March in old Massachusetts." Snow fell from mid-December until New Year's Day, piling one to two feet deep in the Willamette Valley and three to four feet deep in the mountains. "The water swept about one half of Linn City clean & the upper end of Oregon City suffered materially," Ives told his brother. "Dr. McLaughlin's [sic] Sawmill & the bridges, flumes, races etc. were all taken away."[1] The first week in January, warm rains dissolved the snow to flood most of the low-lying prairies and towns.[2]

Settlers neared a state of famine as winter wore on and their cattle and horses died of starvation and exposure in the cold rains. "The last seasons emigration is really suffering," wrote Butler Ives. "They have lost considerable of their stock & provisions are so high that many of them are unable to purchase . . . A great many of the old settlers have been living upon allowances of flour & some on meat, there is no fat beef in the country." Oats and wheat had been scarce during the Fall, and much of what was grown had been shipped to supply California's lucrative market. "It would be a great blessing if it affected only the old settlers," Ives quipped, revealing his Yankee prejudice against the seeming improvidence of emigrants from Southern states, "but as it is, it has learnt them that hereafter they must make some provisions for the winter, both for their stock & themselves."[3]

Holding new subdivision contracts, six deputy surveyors fanned out through the central Willamette Valley to fill in the squares on Preston's checkerboard. Butler Ives paced sodden ground along the Willamette and the Long Tom rivers near present-day Monroe, Junction City, and Veneta, while James Freeman worked a tier of five townships across the valley between Muddy Creek and Sweet Home. George Hyde and his company surveyed between Lebanon and Philomath while Joseph Hunt subdivided small valleys in the Coast Range foothills west of Forest Grove, Yamhill, and McMinnville. Robert Elder's contract lay just south of Hunt's in the western Willamette Valley from present-day Amity south to Corvallis.

In the Calapooia River Valley east of present Brownsville, Kimball Webster pressed on in pounding rain.[4] Supervising his first solo contract, the deputy surveyor contended with David McLellan, an assistant who complained at every hard task and constantly threatened to quit. The deputy considered firing the whining troublemaker but hesitated because McLellan would take his friend with him. One day as the men waded in deep, cold water, McLellan again threatened to leave. Exasperated, the deputy surveyor told him to either be quiet or quit. As Webster expected, McLellan left with his companion, leaving Webster with too few men and a half-completed contract.[5] Disruptions to the survey crews were an inconvenient reality for the surveyors. Illness, injury, incompatibility, or the incessant grousing of malcontents like McLellan repeatedly plagued the deputies. This time, Webster was fortunate to find replacements in Albany — Benjamin Shreve and C. G. Abbey — "two men that proved themselves to be very good help."[6]

On New Year's Day, in a cold wind, Butler Ives chained his line seven miles over the prairie. Detained by high water at Marysville for a week, it had taken the company over two weeks to reach the survey location west of Eugene. When they finally reached their destination, the men stayed several days with settler Prior (or Pryor) F. Blair. Blair and his wife Eleanor had settled their donation land claim in the late 1840s in what is now the Whiteaker neighborhood of Eugene. In addition to farming, Blair operated a sawmill on his claim. Ives and his men stayed a week with the settler, building a boat while they waited for their campman to arrive with tents and provisions.[7]

Ives and his crew pushed on through a wet, miserable January. The company narrowly escaped tragedy when McFall, balancing across the Willamette River on a log, tripped on a limb and fell in. The current quickly sucked him under and, as he grabbed for overhanging branches, he dropped the compass and tripod in the fast-moving river. Keen and Potter scrambled out on the log and, one holding on to the other, twisted under the water to free their trapped colleague. Relieved that McFall was safe, but frustrated at the loss of equipment, Butler Ives went to Oregon City for another instrument. In the meantime, the remorseful McFall searched the icy waters until he found the tripod. He waded back into the cold river the next day to look for the compass, triumphantly pulling it out only slightly damaged. Equipment in hand, the company resumed the survey rather than wait for Ives' return.[8] Unaware that McFall had found the instrument in usable condition, Ives bought another solar compass from draftsman DuBois for $250.

Before rejoining his men, Butler Ives wrote his brother from Oregon City: "It is a bad hit to have to lose the good weather & let the rest get the start of me when they have the best work & only half as far to go to get it."[9] The surveyors' profits depended on speed. Delay, whether due to rugged terrain, twisting rivers, or extensive travel, cost time and money. The more remote a survey from Oregon City, the higher the deputy's overhead for travel and food expenses. Ives' complaint revealed his growing resentment of Preston's granting Freeman, Hyde, and Elder the best survey contracts. Of the contracts issued out that winter, Ives' difficult townships lay the farthest south. "I have the Willamette river to survey through two townships yet," he grumbled. "The bottoms are the worst there is . . . being about 2 miles wide & very low, all cut up with channels sloughs etc. & I am compelled to leave them at every slight rise of the river . . . The high price of provisions & my usual luck will take off some of the profits."[10]

Despite his challenging assignment, Butler Ives held hope that his professional situation might improve. Oregonians could not vote in the presidential election of December, 1852, but Ives, who watched the Territory's newspapers for results, welcomed Franklin Pierce's victory.[11] The impending inauguration surely signaled the replacement of Oregon's surveyor general and some new policies. "If Preston does give Hyde, Freeman, & his brother etc. their choice in the next contract as he has promised them he would," Ives told his brother, "he has an eye on the results of the fourth of March & I accidentally found out that he advised his friends to 'make hay when the sun shines.' It is likely that Hunt & myself will get work enough to do, but the Columbia bottoms will be a specimen of it Well let them go it when they can for by and by, they *can't*."[12]

"Politics are of about the same stripe as when you left, only some Whig officers are thinking that after all 'there is no place like home,'" Ives told his brother. "The opinion here is that Judge Pratt will come back with a Surveyor General's commission in his pocket, but there is no knowing what President Pierce will do."[13] Of his various colleagues, Ives reported that George Belden was now chief clerk, while Wells Lake was temporarily in the lumbering business. Although DuBois was still a draftsman in the office, he shared in a mercantile business at Albany. Hunt and the other surveyors were fine. Unfortunately, Preston would not receive official approval of William Ives' field notes for at least another month, and it would take Butler Ives until the first of April to send the remainder of his brother's payment.[14]

～

In early 1853, John Preston reported to a new man in the nation's capital. John S. Wilson, Acting Commissioner of the General Land Office, had replaced Justin Butterfield as Commissioner. Wilson, who had supervised much of the office's business during Butterfield's term, immediately challenged Preston to defend his interpretation of the donation land laws. Political pressure on the Land Office had mounted when Territorial Representative Joseph Lane publicized a scathing letter ostensibly written by an Oregon citizen questioning Preston's interpretation of the donation land law.[15]

By the Act of February 14, 1853 (10 Statute 158), Congress extended the provisions of the 1850 Donation Land Act to December 1, 1855, and allowed settlers to purchase land at $1.25 an acre after two years residence.[16] This legislation encouraged more families to journey west. Although the overland immigration in 1853 would be lower than that of 1852, at 7,500 persons it would substantially increase the Territory's population, both south and north of the Columbia River.[17] President Fillmore signed a bill creating Washington Territory as he left office in March, 1853, but it would be some time before a surveyor general would open a new office there. Until then, Oregon's surveyor general would be responsible for the new Territory that encompassed all the country west of the Continental Divide and north of the Columbia River and the 46th parallel of latitude.[18]

In a political move closer to home, the Territorial House of Representatives backed a council resolution asking the president to move the surveyor general's office to Salem from Oregon City, reasoning that most of the settled lands were legally claimed by people living near Portland and Oregon City, while "two-thirds of the inhabitable portion of the country lies south of Oregon City." Supporters of relocating the office suggested that to force more distant settlers to go to Oregon City on land claim business would cause them "great trouble and traveling expenses." [19] Shallowly submerged in the Democratic legislators' concern for settler convenience was their unwavering effort to divert political power from Oregon City to Salem. Calling it "an incontrovertible fact that the seat of government is the proper place for said office," the legislature sought to strengthen Salem's position as the Territorial capital.

While most emigrants halted their wagons in the Willamette Valley, some rolled south to take lands in the newly opened Umpqua and

Rogue River valleys. Settling claims throughout the western Oregon valleys, the newcomers peppered the surveyor general with questions about rights of dead or absent spouses, property line conflicts, and claim jumpers. From Yamhill County, Elijah Dodson, who had arrived in Oregon in 1847 with his wife and six children, wrote Preston a story of death and loss, "In the winter of 48 I bought a Claim and settled in the next spring. My wife deceased and in the next fall I married My present wife with 8 children who lost her husband on Snake River in 47 on her way to Orrigon. My children was all of age and was all married and settled before the donation law occurred I would like to know who must [be] the owner of half my Clame or wheather My preasant wife can hold a Claim ajoining mine."[20]

Clatsop County resident Preston W. Gillett chose the third person to outline his problem, "A person whose wife was a squaw took a clame here, & after living upon it about 4 months she died, he then sold the clame to a gentleman who is now living upon it. The first owner never had the clame surveyed. Did she perfect a title to half the clame? If so, who will be her heir? as she had no children."[21] Lewis Cline, who had a 640-acre claim in Linn County and an absent neighbor, peevishly inquired of Preston, "What does it require to constitute continued residence and cultivation will it answer the purpose for a single man to merely eat and sleep on his claim occasionally and do little work now and then and spend the rest of his time at his fathers house [?]"[22]

In each of these cases and thousands more, vulnerable settlers clung to their claims, for they had chanced everything on the promise of free land. They owned little of value and cash was scarce on the frontier; land was their only source for shelter and food and greedy neighbors or an unexpected death could suddenly shatter their tentative existence. Ellen Smith's letter from Marion County showed stark fear. "Mr. Wright and myself have tried to settle about the peace of Land that he proved up of mine and he will not settle with me attall and Sir I appeal to you to know what to doe as I am a Lone woman I wish you to write me how to precede in the Case and I will take your advise and doe as you say and write as soon as you can if you don't I shall lose the Land as I am a Lone woman and have no person to aide me."[23]

Spring, 1853

In April, Preston had received enough funds from Washington to award new survey contracts. As Butler Ives anticipated, the surveyor general gave him four boggy townships in the Columbia River bottoms from Sauvie Island to Deer Island, as well as the broad plain that stretched from Vancouver north to the Lewis River. His contract also covered surveys of the arable lands along the Toutle River, the broad prairies of the Cowlitz River Valleys, and the valleys of the Newaukum River and Chehalis River, all in Washington Territory.[24] And, as Ives had also expected, George Hyde won the coveted contract to extend the Willamette Meridian north of the point reached by William Ives at Puget Sound in 1851. Surveyor General Preston named his own nineteen-year-old brother, Josiah Walker Preston, as Hyde's co-surveyor. Born on August 31, 1833, in Warsaw, New York, the young man had recently arrived in Oregon. [25]

At the same time he awarded the public land survey contracts, Preston hired several men to survey donation land claims in the Willamette Valley including Joseph Latshaw, Israel Mitchell, Jerome Green, and Daniel Murphy. Other recruits were Lafayette Cartee, a member of the Territorial House and Oregon City engineer; Anson Henry, Yamhill County resident and former Oregon Indian agent; and the mathematically gifted Harvey Gordon, who had settled in Oregon seven years earlier.[26] Other men who joined Preston's team: John Trutch, a skilled engineer, had come to Oregon in 1851 to join his brother, Joseph; Nathaniel Ford had arrived in Oregon from Missouri in 1844 and settled in Polk County; Pennsylvanian Robert Short had arrived early in Oregon and secured a claim in Clackamas County; Medders Vanderpool had come to Oregon in 1846 from Missouri and settled a donation land claim; and England-born surveyor Samuel D. Snowden had a land claim in Yamhill County.[27]

One highly skilled deputy surveyor's name was missing from Preston's roster for the spring round of public land survey contracts. James Freeman left Oregon in the late spring for San Francisco. Preston's impending dismissal may have contributed to Freeman's decision, and it was well known that California Surveyor General John C. Hays paid his deputies more than they could expect in Oregon. As one historian observed, "The Federal Government was not willing to pay .ⁱ. as much in Oregon as it would pay for like services in California. The surveys in the latter State were pushed much more rapidly, the allowances . . . being three, four or five times greater than for Oregon."[28] In Freeman's departure, Preston lost the second of his original lead surveyors. At twenty-three, Butler Ives was now the most experienced, followed by Joseph Hunt and George Hyde.

∽

On April 18, Butler Ives, Joseph Hunt, and a crew of six headed into the field for an extended tour on Contract 32. At Linnton on the Willamette River, the company shouldered packs and hiked into the mountains. From the last week in April until late May they endured the swampy lowlands along the Columbia River until high water stalled their survey. Abandoning work there temporarily, the men pushed farther north in Washington Territory.[29] They boated down the Columbia to Monticello and started up the Cowlitz River. On June 1, after three days of hard rowing, they reached Warbassport, where Butler Ives had visited nearly two years earlier. Ferrying across the river, the men strapped on their packs and hiked into the woods.[30] Ives and Hunt surveyed township boundaries as a team during June and then divided the crew into two separate parties to efficiently complete the subdivisions (Figure 43).[31]

Figure 43 (at right). Survey Contract 32: Townships 2–5 North, Range 1 West and Townships 9–12 North, Ranges 1–2 West, W. M.; Butler Ives and Joseph Hunt, 1853. (Map image by Allan Cartography, Medford, Oregon, 2008.)

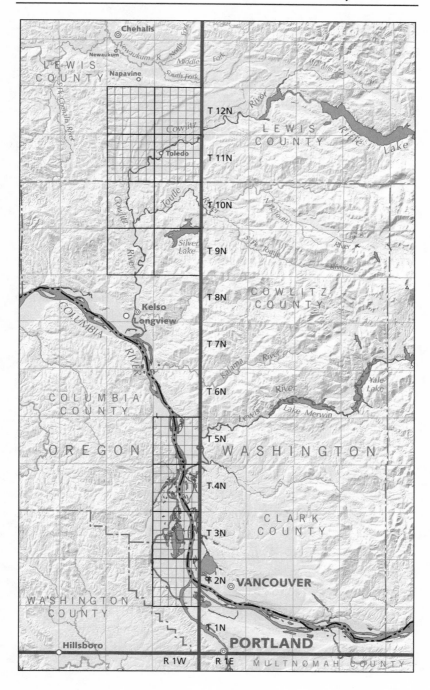

Although Ives and Hunt leave no written record of personal encounters with donation land claimants on this survey, in one instance the settler documented an exchange. In early August, Joseph Hunt and his men chained across prairies southeast of Chehalis, Washington, and stopped to rest on Henry and Maria Cutting's 326-acre claim. A cousin of Hudson's Bay agent George B. Roberts, Maria Cutting had arrived in Oregon with her husband in September, 1852, and settled on "Cutting Prairie" in the Napavine vicinity. On this mild August evening, Mrs. Cutting looked out her door to see Hunt's company camped a short distance from the house. When Henry Cutting walked out and invited the group back for supper, she momentarily panicked. "I had nothing but bread and Butter a Small piece of meat and Onions, which I felt ashamed to put on the table fearing they might be offencive, the Onions I mean. But indeed they liked them. The next day we went and Hunted up a few Blackberries and made a pie which they thought Delicious, so we got along all right."[32]

Miles to the north of Ives and Hunt, George Hyde, Josiah Preston, and their seven-man crew began their contract to extend the Willamette Meridian north from the monument William Ives had built on the south shore of Nisqually Reach on Puget Sound two years earlier.[33] Hyde and Preston entered the Sound, surveyed their line over the westerly tip of Treble Point on Anderson Island, across Drayton Passage, along the east side of the peninsula dividing Case and Carr Inlets, and ended the meridian survey on the north line of Township 21 at present-day Glencove (Figure 1).[34] When *Oregon Statesman* editor Asahel Bush learned of George Hyde's plum assignment in Washington Territory, he publicly accused Preston of the same injustice that Butler Ives had described privately to William Ives: "The Surveyor General has a 'brother-in-law' who has had the best contracts for Surveying ever since he has been in the office . . . [and] just upon the eve of being turned out [has] . . . let him a lucrative contract which will employ him the greater part of the summer . . . I am informed that Mr. Preston, knowing that he will be speedily removed, has let to his favorites contracts to survey all the best portions of the country."[35]

At the same time that Bush was skewering Preston for distributing survey contracts unfairly, Preston's Whig supporters were promoting the surveyor general as a candidate for Territorial delegate to Congress. The newspaper editor declared that Preston's followers had "gone crazy with the opinion that [he] possesses a popularity which will carry things by storm," suggesting that they may as well "be undeceived now as ever."[36] To discredit the surveyor general's candidacy, Bush printed a letter under the pseudonym "Candor" accusing Preston of charging unlawful fees while accepting a large government salary "for which he requires the toilsome settler to pay from his scanty means." For his own part, "Candor" said, "I paid it rather than have a difficulty, as I was convinced that the power was all in his hands, and I must submit to his demands . . . or lose my land." The writer closed by warning readers that, if elected to Congress, Preston would likely charge extra "for every service rendered to his constituents."[37]

By late spring, Franklin Pierce had removed most Whig officials in Oregon, forcing them, as the *Statesman* put it, "to walk the plank." The new president named Joseph Lane as governor, George L. Curry as territorial secretary, and George H. Williams as chief justice. Lane, who replaced Gaines on May 16, served only three days in office before resigning to continue as Oregon's representative to Congress. Appointed as acting governor, George Law Curry would serve through December 2, 1853, when John Wesley Davis arrived to take over the office.[38] Pierce ignored Preston while purging his first batch of Oregon Whigs, however, and, for the time being, the surveyor general hung on to his job.

Prudently diversifying his financial interests, Preston joined Sidney Moss and other Oregon City entrepreneurs to organize the Rockville Canal Company. The group planned to build a canal and breakwater at Willamette Falls, transport freight, and promote manufacturing.[39] Preston also joined Oregon resident James O'Neill and his brother, Daniel O'Neill, in the mercantile business. The former, a native of New York State, had recently arrived in Oregon, while his brother ran a general store in Oregon City. The three investors built

Figure 44. John McLoughlin *(center)* with his granddaughters, Margaret Rae Wyant and Louisa Rae Myrick, and, in back, George Hyde *(left)* and Daniel O'Neill *(right)*. (Oregon Historical Society, OrHi 45701.)

their store on Main Street just north of former Provisional Governor George Abernethy's building, where Abernethy now ran his successful mercantile business. Preston and the O'Neills advertised wholesale and retail sales for a wide selection of items including building supplies, tools, saddles and bridles, oil, salt, and sugar.[40] Firmly established in business and still drawing his government salary, John Preston figured prominently in Oregon City's social circle. He and Lucy Preston welcomed Governor Gaines, John McLoughlin, and other notable Oregonians to their home, contributing to what Elizabeth Millar, one of the teachers escorted by the Preston's to Oregon City, described as the "gay and fashionable air . . . imparted to society . . . by the families of the territorial officers " (Figure 44).[41]

If a "gay and fashionable air" permeated the atmosphere as socially prominent Oregonians entertained each other at lavish parties, so did the winds of change. Two years had passed since Preston and his cadre of surveyors began the public land surveys. Now, national and regional politicians compromised Preston's reputation; thousands of new settlers clogged the clerical bottleneck in the surveyor general's office, and two of the Territory's most able surveyors, William Ives and James Freeman, had left the country. Uncertainty lay ahead for the remaining deputies who would soon work under a new surveyor general. The number of profitable surveys dwindled, while more surveyors competed for work and late appropriations from Washington threatened to delay payments for everyone. Now old hands in Oregon, Butler Ives, George Hyde, and Joseph Hunt took it one contract at a time.

Chapter 8 – 1853

Removal and Replacement

"When will you send sum one to survey?"

As the summer of 1853 unfolded, Whigs and Democrats battled over the remnants of political power left in Oregon City. Democrat legislators accused Preston, who was still based in the former capital, of delaying Territorial business when it suited his purpose. In one skirmish with Democrats, the surveyor general scuffled with judicial officials when the Supreme Court term opened in Portland. Ralph Wilcox, a Democrat, sat as clerk of the court in the absence of A. P. Millar, Preston's former clerk, who had left for the East Coast earlier in the year.[1] When Allen Seymour, Millar's deputy clerk and a part-time clerk in the surveyor general's office, refused to turn court documents over to a court messenger to take to Wilcox, the court declared Seymour in contempt.

The *Oregonian* and the *Statesman* — the latter paper had relocated to Salem in June — jumped at the chance to fuel the partisan fray. Preston informed the court that, for Millar's security, he had advised Seymour to give the papers *only* to the clerk of the court or another authorized person and to get a receipt.[2] The *Statesman* claimed that court officials knew Millar was gone and thought that Seymour had also left for the East after bundling all the court papers into packages to leave in Preston's care. Editor Bush branded Seymour a tool in the

145

hands of the Whigs, and accused Preston of refusing to recognize Democrat Wilcox as legitimate clerk of the Supreme Court. Officials jailed Seymour who, after spending some time behind bars, accompanied the Marshal to the surveyor general's office, retrieved the papers, and handed them over.[3]

On June 30, 1853, President Pierce fired John Preston.[4] "There is no longer any doubt as to the removal of the present Surveyor General," the *Oregon Spectator* editor announced when reports of Preston's long-anticipated ouster reached Oregon City several weeks later. Clearly a supporter of the Whig official, the editor complimented him for endearing himself "to everyone with whom he has transacted business."[5] Although news of Preston's expulsion did not surprise his surveyors, they wondered how it would affect their chances for work. From his camp in Marysville, claims contractor Dennis Hathorn sent Preston a letter that got right to the point. "Is it true that some one has been appointed to succeed you as surveyor Gen? & if so will you make any more contracts before your successor arrives?"[6]

In the ensuing weeks, however, Preston stayed on as acting surveyor general. "As the public are asking why it is that the official head of Mr. John B. Preston . . . still remains on his shoulders, every other Whig Surveyor General having been decapitated," the *Oregon Spectator* editor quoted a Washington *Republic* correspondent as saying, "[Preston's] exemption . . . probably results from the fact that . . . Congress has so complicated the duties of his position, making him a Receiver and Register, as well as a Surveyor General, as to render it a rather dangerous experiment to substitute any other gentleman in his place in the present condition of his office."[7]

In midsummer, 1853, Commissioner Wilson of the General Land Office wrote Preston that he had appointed Charles K. Gardner as surveyor general for Oregon Territory.[8] Knowing that his replacement would not arrive for several weeks, Preston remained at his desk in the surveyor general's office. While Preston did not abandon his former post during these "lame duck" months, he devoted increasing time to his private business interests and correspondents had a difficult time contacting him. Jacksonville attorney Thomas McF. Patton wrote Preston from the booming mining town with questions about notifications, adding "I have written to you at divers times but rec'd no answers."[9] Preston and the O'Neills moved their store into John McLoughlin's new building on Main Street and advertised as "retail and commission merchants." Located conveniently near the steamboat landing, the new $10,000 structure housed Preston and O'Neills' business on the first floor and additional rooms on the second level.[10]

Far from Oregon City and John Preston's summer of uncertainty, Robert Elder surveyed townships from Yoncalla south to the vicinity of Roseburg, where Aaron Rose's improvements on Deer Creek formed the center of a growing community. Settlers had poured into the region in the preceding two years, building log houses and barns on grasslands along the North and South Umpqua rivers and establishing the town of Winchester as the county seat on the North Umpqua. Near Yoncalla, Elder watched while workers built Charles Applegate's new home (Figure 45).[11]

When they finished the Umpqua survey in August, Elder and Webster decided to try for a contract in the Rogue River Valley. Just before Elder left to negotiate the project with Preston at Oregon City, the surveyors heard news of bloody hostilities between Indians and white settlers in the Rogue country.[12] Tensions had escalated in December, 1852, with the murder of seven miners at Galice Creek and the subsequent retaliatory killing of six Indians by Jacksonville men. In early August, 1853, Shasta Indians killed a miner and left his body outside Jacksonville; soon after, miners hanged two Shasta men and a nine year old boy. "Volunteer" militia then attacked a Shasta village on upper Bear Creek and an avenging band killed several white men at Willow Springs. After bitter fighting on Evans Creek north of the present-day town of Rogue River, in the last week in August, the region's settlers fiercely opposed proposals for a peace treaty. Jesse Applegate wrote Preston, "The Indian disturbance is daily assuming a

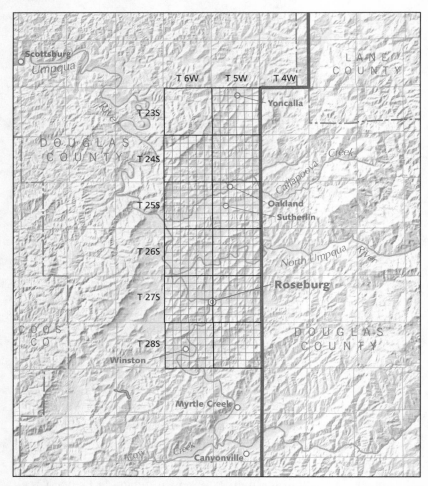

Figure 45. Survey Contract 31: Townships 23–28 South, Ranges 5–6 West, W. M.; Robert Elder and Kimball Webster, 1853. (Map image from Stuart Allan, et al., *Atlas of Oregon.* Eugene, Oregon: University of Oregon Press, 2001.)

more serious character and while armed citizens may be seen rushing in one direction to the seat of war, many of the unarmed miners are fleeing in the other for safety." Caution winning over ambition, Elder and Webster abandoned their plan for the Rogue River Valley and returned to Oregon City.[13] "Owing to the difficulties with Indians in the Rogue River Valley," Preston wrote Commissioner Wilson, "it was found impracticable to extend the Willamette Meridian to the south line of the Territory. This year, no one was found who would undertake the work. It is to be hoped that the surveys in the Rogue River Valley can be prosecuted early next season."[14]

As soon as Elder and Webster left the Umpqua country, Jesse Applegate wrote Preston again, this time to report problems with the surveyors' markers. Assuring the surveyor general that he did not doubt the deputies' honesty and dedication, Applegate nonetheless criticized their work. Robert Elder, he complained, had resurveyed some of his lines and removed the original corner stakes, so that now it was "impossible from the present appearance to tell when the post is missing whether the true place of the corner is at the old or new position."[15] Applegate had bad news, too, about the earthen mounds Elder's crew constructed: "They [have] in most cases been destroyed by cattle. It is a well known fact that that cattle will playfully attack a newly made mound or other elevation and never suffer it to rest until it is leveled with the earth . . . They are also similarly disposed towards posts or stakes and will push them down unless they are planted sufficiently strong to resist their strength."[16]

In a separate letter of the same date, Applegate answered Preston's recent query about the best route for extending the Willamette Meridian from its present terminus south of Roseburg through the Rogue River Valley to the Oregon-California boundary. "The Umpqua Mountains at which Mr. Freeman terminated the East Boundary of Range 5 West is of so rugged and steep a character," Applegate advised Preston, "that the true horizontal distance on a straight line across the range cannot be accurately ascertained by the ordinary mode of measurement." Describing iron-laced rock that foiled the magnetic compass and dense forests that rendered the solar compasses useless, Applegate declared that to extend the surveys across the steep ridges between the Umpqua and Rogue valleys would be impractical.

By climbing high in the mountains a few miles east of Freeman's ending point on the meridian, Applegate said, one could see off to the southeast "a high conical peak snow clad like Mt. Hood . . . This cone erroneously called Mt. Pit on the maps is East and in full view of the Rogue river valley." He recommended using this mountain, present-day Mount McLoughlin, as an efficient way to link the public land survey to the Rogue country by triangulating "from a base connected by course and measurement with Mr. Freeman's line in the Umpqua valley." A surveyor could determine the mountain's position in relation to the Willamette Meridian and later measure from another point in the Rogue River Valley. "That valley's position from the mountain, and hence from the Willamette Meridian could also be ascertained," Applegate declared. It would be easy, he said, "to establish some Township corner in its proper position in relation to the Base line and Meridian, and from it extend the public surveys in all directions until all lands fit for settlement were included."[17] Knowing that the township surveys in southern Oregon would fall to his successor, Preston remained non-committal on the subject. Having given Preston his best advice, Applegate packed his new solar compass and joined Brevet-Major Benjamin Alvord to survey a road from Myrtle Creek to the Rogue River Valley.[18]

In September, Preston awarded three final survey contracts for the year, his last as surveyor general. Short of surveyors since Elder had returned East and Butler Ives, Joseph Hunt, and George Hyde were all in Washington Territory, Preston hired two more men. Daniel Murphy, a native of Ireland and settler in the Champoeg area of Marion County, took Contract 36 to survey west of the Willamette River in the small, fingered valleys of the Coast Range along the Luckiamute River, Mill Creek, and Rickreall Creek. Anson Henry, who had worked as a claims surveyor and was now a member of the territorial legislature, subdivided townships west of Portland in the Tualatin Hills and in the area of the Chehalem Mountains north of present-day Newberg, Oregon.[19] When Preston offered Kimball Webster his pick of any contract remaining in the areas left to be surveyed, Webster chose townships in the Long Tom River Valley and smaller valleys in the rugged foothills of the Coast Range.[20]

Fall, 1853

Butler Ives and Joseph Hunt left the Cowlitz country on August 24 and returned to the Columbia where they had temporarily stopped work several months earlier. Ives ran township lines with McFall, Charles Terry, Erastus Smith, and Sewell Truax, while Hunt subdivided with Lyman Chittenden, William West, William Gibson, and William Byers.[21] The sun shone throughout September, but in October the weather turned bad. As McFall observed, "Had a great deal of rain and cold south wind, which made it very disagreeable to work in the low bottoms."[22]

When the surveyors got back to Oregon City in late November, they found Charles Gardner sitting at Preston's desk. Gardner had arrived by steamer at the first of November, accompanied by his sons Charles T. and George Clinton Gardner, who would work for their father as deputy surveyors.[23] Charles Kitchel Gardner was sixty-six years old when President Pierce named him to be Oregon's second surveyor general. Born in New Jersey in 1787, he rose rapidly in the US Army to become captain of the Third Artillery during the War of 1812 and adjutant general of the division of the north in 1816. Resigning his commission in 1818, he filled several government positions: Assistant Postmaster General from 1829–1837; Auditor of the Post Office Department, 1837–1841; Commissioner to Settle Affairs in Connection with the Indians of the Southern States, 1841–1845, and Postmaster of Washington, DC, 1845–1849 (Figure 46).[24]

Asahel Bush gave Gardner a cautious nod two weeks later, while castigating Preston for illegally charging settlers fees to file their notifications. The new surveyor general, on the other hand, Bush wrote, "concludes that he has no right to fees, and will in no case charge any. This first act of his administration will commend him favorably to the people."[25] Gardner's deputy surveyors treated him courteously enough, but Kimball Webster had reservations. "Colonel Gardiner," he said, "[is] a good man in his place, but he [knows] very little in regard to the public land surveys. As Mr. Preston was a practical engineer and a

Figure 46. Charles K. Gardner, Oregon Surveyor General. (Reprinted from Benson J. Lossing, *The Pictorial Field-Book of the War of 1812.* New York, New York: Harper & Brothers Publishers, 1868.)

surveyor, it [is] a poor exchange in a practical sense."[26] For John Preston, his removal as surveyor general was far overshadowed by devastating loss when his twelve-year old daughter died on November 24. Reverend George Atkinson called on the suffering parents, "Last night . . . Clara Preston . . . died of croup and congestion of the lungs. It was a house of grief. This is the last of three children. They are written childless."[27]

To distract himself from his sorrow, Preston returned to work. A civil engineer, businessman, and an attorney, the former surveyor general eyed a new opportunity in his successor's workload. The two men had been acquainted in the East, and Gardner had sought an appointment for one of his sons on the surveys in Oregon prior to Preston's departure for Oregon. As Gardner settled into his post, Preston opened a private land law business in an office over his store to "prepare notifications and Proofs for securing Donation Rights, prosecute cases of conflict of boundary before the Surveyor General and the Departments at Washington and attend to any business pertaining to land, or Land Laws, in the Territories of Oregon and Washington."[28] The *Statesman* editor scoffed at Preston's suggestion that he merely wished to assist Gardner, alleging that the former surveyor general had originally intended to leave the Territory when his replacement arrived. Instead, the acerbic Bush declared, "upon deciding that the new appointee was not likely to perform his duties, [Preston] . . . advertised himself ready to perform all kinds of land-office business, but at a double the rates charged when in office."[29]

Surveyor General Gardner had been in Oregon City only a few days when he received his first letter from Jesse Applegate. Dismayed by delays in the land claim surveys and frustrated by Preston's lack of enthusiasm for his ideas, Applegate wrote Gardner, "In the populous county of Jackson *not a line has been run*; in the counties of Umpqua and Douglas only a few Townships have been subdivided." Even though the public surveys might reach into the region in 1854, Applegate protested the former surveyor general's ironclad requirement that settlers have their land surveyed first for their notification and later by the government surveyor. Applegate asserted that the law "*already* provides that each claimant may have his claim surveyed by the proper authority prior to notification and upon the completion of two or four years . . . he may have his final certificate and patent." Were Gardner to support the latter scenario, Applegate declared, "The people will cease to regret the slow progress of the surveys, and to besiege the halls of legislation for redress."[30] As if to underscore Applegate's assertion, Abraham Smith wrote Gardner on November 20 from Calapooia, desperation seeping from the page:

Will you Please Send sum one to Survey Township 14 South Range Three West whare as there was an order sente and was signed By Mr Mikels Parkes and Keeny And others and there is a number others in this naberhood that have Proved up sense. They wantes there claims run. There is William Hawk, P . . . White William Smith and Sum and that livs on fracksons and can't prove up until lines is run. I Abraham Smith was down to prove up but could not till James Keeny and Parks had there lines run . . . pleese rite When will you send sum one to survey?[31]

~

At the close of 1853, Charles Gardner still had plenty of contracts to oversee. Many of the areas left for survey were located in the foothills of the Cascades or the Coast Range, where settlers farmed small, fingered valleys. Two large contracts remained, both in southwest Oregon. The first centered in Jackson County's Bear Creek Valley, the second in the other portions of the Rogue River Valley and in the Illinois River Valley. Paperwork piled high in Gardner's office, inundating the seven clerks and draftsmen with applications and correspondence. Cold political and financial reality soon engulfed the new surveyor general as it had his predecessor. Despite his initial positive review, Asahel Bush now scrutinized Gardner intently. Settlers in the southern reaches of the Territory clamored for surveys. Congressional appropriations arrived slowly from Washington, if at all, while Commissioner Wilson shipped regulations and criticism equally in his communications. Charles Gardner had all he could handle.

Chapter 9 – 1854

Meridian South

"To run the Meridian to our southern boundary."

At 6:40 A.M. on April 8, 1854, as passengers boarded the *Gazelle* at Canemah, just upstream of Oregon City, for the run up the Willamette River, the steamer's boiler exploded. The blast ripped open the deck and slammed iron and wood fragments into screaming passengers. Scalding water spewed over the vessel, searing the wounded, many of whom had severed legs and broken ribs. Of the fifty passengers and crew, twenty people died and twenty-five more suffered serious injuries. Surveyor Joseph Hunt, aboard ship to begin a survey contract, died instantly. Late that same day, Gardner, whose son Charles was standing next to Hunt, wrote Commissioner Wilson, "They both had but two minutes stept on the forward deck together, when the explosion took place: my son wonderfully escaped with but slight scalding and a broken arm: Mr. Hunt was cut to pieces."[1]

Steam navigation had begun on the Willamette River late in the spring of 1851, and from then on, shallow-draft steamers transported travelers throughout the valley. The surveyors took steamers to reach contract locations at places near Butteville, Champoeg, and Salem. When rivers ran high, they could travel as far as Marysville and beyond. Although steamers were more commonly damaged by rocks or snags hidden under the water, serious accidents like the one that cost

Joseph Hunt his life happened from time to time. The previous summer, one passenger had died when the boiler of the *Canemah* exploded as the steamer lay near Champoeg. In another accident, the *Fenix* and the *Oregon* collided near Butteville.

Hunt's death stunned Butler Ives. Colleagues and friends of long standing, the two had recently finished the six-month survey contract in Washington Territory together and after William Ives' marriage to Sarah Hyde the previous year — Sarah's brother had married Hunt's sister — they called themselves kin. Ives petitioned the Clackamas County probate judge to appoint John Preston administrator of the estate and gave authorities the names of Hunt's Michigan family. Then he set about the sad task of sorting his friend's belongings. Most of what the deputy owned was on the *Gazelle* — tent, blankets, compasses, his buckskin suit, wool shirts, and gold watch. Ives set aside Hunt's leather trunk, extra clothing, and shotgun, and went to Preston's office to arrange for his friend's burial.[2]

Jesse Applegate gave Surveyor General Gardner a few weeks in office before advising him on the best way to extend the Willamette Meridian through the Umpqua Mountains into the Rogue River Valley. Writing in late January, 1854, he repeated what he had told Preston: surveying the line through "the Canon," the tortuously steep, rugged, country along Canyon Creek north of present-day Canyonville, was impractical, and Gardner should triangulate measurements to the High Cascades' Mount Pitt (Mount McLoughlin). Turned down flat by Preston for contracts, Applegate asked the new surveyor general to hire him for the meridian extension and Rogue River Valley surveys.[3] As it turned out, Applegate's letter reached Charles Gardner well after January 4, the day he awarded Contact 39. Butler Ives and George Hyde approached the surveyor general as a team and presented bonds guaranteed by John McLoughlin, John Preston, and Eugene F. Skinner of Lane County and won the coveted assignment.[4]

156

A little more than a week later, Gardner defended his selection of surveyors to Governor John W. Davis, appointed by President Pierce in late 1853, who favored Lafayette Cartee for the contract. Cartee was born in New York in 1823 and entered St. John's College in Cincinnati, Ohio, in 1846. He later served as an instructor of mathematics and civil engineering. Cartee came by ship to San Francisco in 1849 and later that year opened an engineering office in Oregon City. Politics played a not-too-subtle role in Davis' support for Democrat Cartee, a representative to the territorial legislature from Clackamas County.[5] "To run the Meridian to our southern boundary and the exterior lines of townships connected with it," Gardner wrote Davis, "[requires] surveyors experienced in the use of Burt's Solar compass; and I have fixed upon two . . . of the best who remain in this Territory (since Mr. Freeman has joined Col. Hays in Calif.)." To further convince the Governor, Gardner added, "I am, besides, well informed that they are both Democrats."[6]

Contract 39 included extending the Willamette Meridian from the point near Canyonville, where James Freeman had stopped more than two years earlier, south to the Oregon-California line, as well as the survey of several townships. Lacking information about the Rogue country, Gardner did not specify townships for survey, instructing Ives and Hyde instead to survey boundaries for twenty-three townships and to subdivide ten of those townships. Leaving the final selection up to the two deputies, he set their pay at $25 per mile for the meridian and necessary offsets, $17 per mile for standard parallel and township lines, and $11 a mile for the subdivision lines.[7]

Three years had passed since the first few settlers had arrived in the Rogue River Valley late in 1851, and after the heavy emigrations in 1852 and 1853, farmers had most of the region's arable lands under claim. With the public land surveys long overdue and unable to gain title to their land, the valley's settlers wrote Gardner complaining of disputed claim lines. W. H. Smith started at the beginning to relate his complex situation:

In the first place there was a man by the name of Wills took the said Claim up one year ago this spring. Held the claim about two months And sold it. He never lived on the Claime put No improvements on it whatever And in the mean time he bought the claim back again held the claime as was stated afore untill last fall. During the time of the break of the indians he got killed the claim laid vacant until the emigration came in. in the mean time it was jumpt the Man that jumped it was [run] off the next day After he was put off it was jumped by another Man he was sued and beat out of it And he left the claim . . . the said Willses brother came in from of the plains And claimed his right of heirship to the claim. He sued and beat the man out of it that jumped it the second time The deseased Mr. Willses brother after gaining the suit by an unfair trial was put in possession of the claim but never settled on it he went off in to the mines and left the Claim laying vacant paying no attention whatever. Never put any improvements And never lived nor slept on it.

After this time was up I jumped the said Claime They are now trying to [put] me off. The deseased Willses brother has Never been on the Claime but once since he went into the mines. I was on the Claim at the time he ordered me off the claim but I did not Leave it.[8]

Smith said that he was still on the claim and intended to stay there until he heard from Gardner. The surveyor general gave him little reassurance. "The claim of Wills who was killed in fighting the Indians can be secured to his heir or heirs by their proving 'residence & cultivation' only to the day of his death," he told Smith. "If Will's brother beat the men who jumped the claim in a legal trial, I should suppose he could beat you."[9]

At the same time that Rogue River Valley residents pleaded for the public land survey, settlers in other areas of Oregon protested what they considered outrageous fees to process their claim applications. One detractor wrote Commissioner Wilson in Washington to protest Gardner's refusal to assist settlers with their paperwork. "He forces claimants to the expense of feeing lawyers . . . for the purpose of filling up their blanks, when it is evidently the spirit and intent of the law, that there should be no expense about the matter."[10] Politically sensitive to public complaints and eager to deflect blame from his own office, Commissioner Wilson chastised Gardner: "The object of the General Land Office in furnishing printed forms of all documents . . . required in settling claims in Oregon, and in providing sufficient clerk hire, was . . . to relieve the settlers from every expense . . . You will, therefore, extend to them every facility in your power for this purpose, and, if the means placed at your disposal are insufficient . . . you will immediately report the fact, that further appropriations may be asked of Congress."[11]

From Salem, the *Oregon Statesman* editor linked Preston and Gardner in his assault. He first chastised Preston: "The day of official filching and oppression is past, but there still remains the sad reality that *twenty-five thousand hard earned dollars* have been extorted by the last Surveyor General, in the shape of unlawful fees taken from the farmers of Oregon." Targeting Gardner, the editor charged that although he did not charge extra fees, he "declined to do the full duties, and claimants were compelled to resort to the Ex whig S.G., at a greater expense than before, to get their applications to the office, and proofs put in form."[12]

⤚

Ives and Hyde waited January through March of 1854 for snows to recede in the Rogue-Umpqua Divide and for the hostilities between Indians and white settlers in southwest Oregon to subside. Although Joel Palmer and tribal representatives had signed treaties in the Rogue and Umpqua country the previous September, conditions in the region remained tense. The Rogue bands had relinquished their former territory to the US government in return for 100 square miles of reserve and now languished in a confined area that reached up Evans Creek, across the mountains to Upper Table Rock, south to the Rogue River, and downriver on the north bank to the mouth of Evans Creek.

New fighting had threatened to erupt after whites hanged two Indian men at Jacksonville for the murder of a miner the previous year.

The first of April, Ives and Hyde readied for the Rogue country, packing tents, blankets, food, compasses, and transits. On April 6, George McFall and Sewall Truax went ahead, planning to meet the company later on. On April 8, the day of the *Gazelle* disaster, the two reached Albany, where they learned of the steamer's explosion. Delayed a week by the tragedy, Ives, Hyde, and the rest of the company arrived in Albany on April 18. Camped near town, the men spent the next day loading packs, only to be delayed again when a mule strayed away.[13]

Early on the morning of April 21, the company left town in midmorning, men and mules stretched out along the trail, "all armed and equipped and in fine spirits for Rogue river." Twelve assistants joined Ives and Hyde: Sylvester Cannon, E. G. Cowne, Josiah Case, John D. Griffiths, James D. Price, Joseph M. Addington, Edward R. Shunk, George Hadlock, Henry Merrill, John N. Lewis, George McFall, and Sewall Truax. Cannon, Price, and McFall had worked on survey crews for William and Butler Ives, while the twenty-four-year-old Truax, a Canadian native trained as a civil engineer, assisted Butler Ives and Joseph Hunt the previous season in Washington Territory.[14] Cannon, Price, Cowne, and Shunk signed on as chainmen, Lewis and Hadlock as axemen, and Addington and Griffiths as mound-builders.

The first night out the men camped just south of Burlington, a small community east of the Willamette River, about a mile north of present-day Peoria, Oregon. Over the next eight days the company made over 115 miles along the territorial road, traveling as much as twenty-two miles on a good day and as few as ten in driving rains. After waiting out a storm on April 29, the company left the vicinity of present-day South Umpqua rest area on Interstate 5 at noon, climbing fourteen miles into the mountains to Canyonville. On May 1, Ives sent the packers and mules south along the road and, with the rest of the company, shouldered their packs and hiked east into the mountains.[15]

Willamette Meridian South, May 2 to14, Thirty-One Miles

Inching up the steep slopes, Butler Ives calculated the distance and direction toward his goal and, two and one-half miles into the hills, claimed the treasure: Freeman's now-weathered wooden stake. Inscribing his leather-bound journal, "Linear Field Book of Willamette Meridian," Ives wrote his name and the date, May 1, 1854. "Extension of the Willamette Meridian on 2 offset commensing at a post on West boundary of Sec. 30 Township 30 South, Range 4 West of the Wm. Meridian, as established by James E Freeman on the 179 mile" (Figure 1).

On the morning of May 2, the company rose at dawn. At the front of his field book, Ives charted the month's chart of solar declination, the sun's angular distance north or south of the celestial equator, with readings taken at 6 A.M., 9 A.M., noon, 3 P.M., and 6 P.M.; in another volume he kept the geodetic notes.[16] The men chained the meridian three tortuous miles to slide into a ravine so narrow and steep that they had to build a platform to sleep on. The following day they made another three miles where "in the whole distance didn't have three-hundred feet of level land."[17]

On May 4, Hyde, McFall, and Hadlock hiked out to the main road to meet their packers and reprovision. Afterwards, the company camped two days on Cow Creek waiting for the weather to improve. On May 8, the men shouldered their packs and climbed higher into the mountains, where north-facing slopes held deep patches of snow. They surveyed the meridian over rugged ground east of Canyon Creek Pass, across Upper Cow Creek and Starveout Creek, chopping through thick brambles to reach the headwaters of Quines Creek southeast of present-day Azalea, Oregon.[18]

On May 10, Ives and Hyde turned west six miles on a third offset of the Willamette Meridian (Figure 1). Stretching west from the Quines Creek headwaters to near the present Interstate 5 rest stop at Cow Creek, the offset avoided the most rugged section of the Umpqua Mountains where peaks ranged as high as 5,000 feet. At the end of the sixth mile, the company turned south again. Traveling south about two miles east of the present Interstate 5, they crossed Wolf Creek and Coyote Creek and, climbing through oak, laurel, and manzanita thickets, stood on the ridge dividing the Coyote and Grave creeks drainages. Several miles to the south the surveyors saw "a high conical Peak [Sexton Mountain]" to the west of the Oregon-California road.[19] The company forded Grave Creek — noting "Mr. Bates' House," where William Ives had stayed two years earlier — and passed just east of Mount Sexton's highest point to camp one-half mile south of Jumpoff Joe Creek, on May 13 (Figure 47).[20]

Backed up to a roaring fire that evening, Ives copied his rough field notes for the day, describing land along Cow Creek as "a narrow, rich valley from 12 to 14 miles long and mostly settled" and Wolf, Coyote, and Grave creeks as "small valleys which will admit of settlement."[21] Ives provides a firsthand description of the terrain and vegetation along the meridian: "[the] Line . . . passes over the Umpqua mountains a very rugged broken cross range, between the Cascade and Coast mountains . . . the slopes are generally very steep and stony . . . [and] heavily timbered with fir, laurel and redwood . . . usually a thick undergrowth of laurel, hazel, arrowwood Live Oak, & mansaneta, an evergreen shrub similar to the Laurel."[22] The names Butler Ives assigned southwest Oregon vegetation were those in common use in the mid-nineteenth century. Simple translation makes Ives' "Redwood" not true redwood but probably incense cedar; his "laurel" madrone or Pacific madrone; and his "mansaneta" either green- or white-leaved manzanita.[23] "Poison oak" was poison oak and it gave the men a painful, oozing rash.

Willamette Meridian South Offset, May 14 to 19, Thirty Miles

Six miles south of Jumpoff Joe Creek, Ives and Hyde looked out over the Rogue River Valley. Waiting out cloudy skies on May 15 just north of present-day Grants Pass, the company turned due east across the wooded hills between the Evans Creek and Ward Creek watersheds.[24] In the afternoon of May 17, Ives left Hyde in charge of the

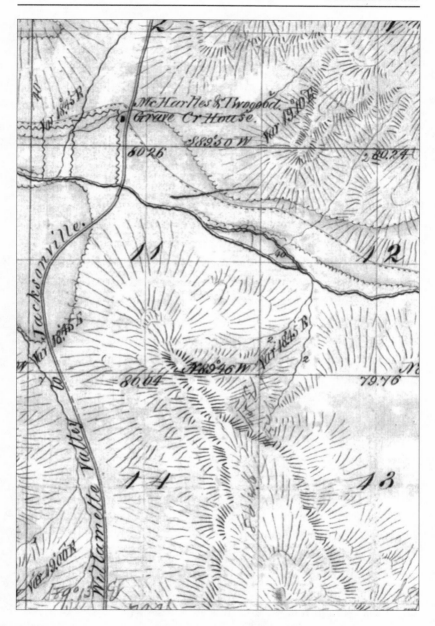

Figure 47. General Land Office Map: Township 34 South, Range 6 West, W. M.; Grave Creek House, Contract 39, Butler Ives and George Hyde, 1854. (USDI Bureau of Land Management, Portland, Oregon.)

survey party while he went after provisions. Ferrying across the Rogue River at Rock Point, Ives led the pack animals along the Oregon-California road. Not far north of the road lay Fort Lane, built in 1853. Like Fort Orford (1851) and Fort Dalles (1852), Fort Lane housed a federal garrison. Headquartered in the compound's log buildings, US Army troops were charged with protecting Takelmas and Shastas from violent civilian "volunteers" and maintaining order in the region.

Leaving the stage road, Ives traveled cross-country to Obadiah B. McFadden's ranch west of Bear Creek near Seven Oaks, northwest of present-day Central Point, Oregon. McFadden, appointed associate justice of the Oregon Supreme Court in 1853 in place of Matthew Deady, had arrived in Jacksonville three months earlier to hold court for the third judicial district of Umpqua, Douglas, Jackson, and Coos counties.[25] From McFadden's ranch, Butler Ives wrote Chief Clerk Wells Lake in Oregon City, "Our worst work is done & I think well, for we have a first rate set of men. The Rogue River Indians are quiet, but [Chief] Tipsey's Band are stealing & threatening some, though we anticipate no trouble from them. There was a man killed by the Indians in the Siskiyou Mountains yesterday. I believed they belonged to a California tribe."[26]

In the spring of 1854, native peoples living on the Table Rock reserve suffered dysentery and food shortages. At the same time, those who had not joined in the treaty and who lived off the reservation committed scattered violence. As their people died around them in the valley, survivors fled into the mountains for safety. The leader Ives called "Tipsey" was Tipsu Tyee, head of a Shasta band on upper Bear Creek. This militant leader refused to attend the treaty negotiations at Table Rock and loathed those who acquiesced to US government stipulations. On the day Ives wrote his letter to Wells Lake, the terrified Shastas of the Klamath River country in California killed Tipsu Tyee rather than join him in a murderous assault on whites.

During Ives' absence, George Hyde pushed east along the offset with John Griffiths assisting him as compassman.[27] Crossing steep country on May 18, the men made two miles before stopping along

Sardine Creek, a tributary of the Rogue River, for their noon meal. One or two of the men had gold pans and soon the entire company knelt beside the stream for some prospecting. "Found several small pieces," George McFall reported, "but not in sufficient quantity to stop working."[28] Moving east, the company reached Sam's Creek near the twin mesas of Lower Table Rock and Upper Table Rock and camped for the night. At sunrise the next day came a heart-stopping moment. "Just as we were getting up," wrote McFall:

> We were visited by a party of 8 or 10 Indians, all armed and equipped and headed by their chief who came to ascertain what our intentions were — whether peaceable or not. We were told by their interpreter that the chief had all he could do to prevent his men from attacking during the night, they thinking we was a war party. But their chief was hoping for the better and wanting to wait until morning when we could have a 'wa wa' (talk) . . . and find out how matters stood. We explained our business and they went off satisfied.[29]

This unsettling encounter underscored the surveyors' vulnerability in the field. Although they made a good-sized company en masse, the men frequently divided into smaller parties for efficiency and speed. Now, McFall noted, there was "a great deal of excitement among the Indians — rather suspicious of our movements."[30] Adrenalin pumping, the shaken men broke camp quickly and moved east along the south boundary of Township 35 South, Range 2 West north of Lower Table Rock. "An Indian Farm lays about ½ mile north," Hyde observed. "There are 4 white men plowing for the Indians & have about 50 acres plowed. This station is on the reserve."[31] The deputy ran the line east three miles and, by the time Ives rejoined the group at midday, approached Upper Table Rock. Prominent mesas of sandstone and basalt rock, Upper Table Rock and Lower Table Rock rise sharply from the valley floor, their horizontal caps topping sheer vertical cliffs (figures 48 and 49).[32]

The men chained the offset up the steep west slope of Upper Table Rock to the base of the cliff where they placed a hatchet as a

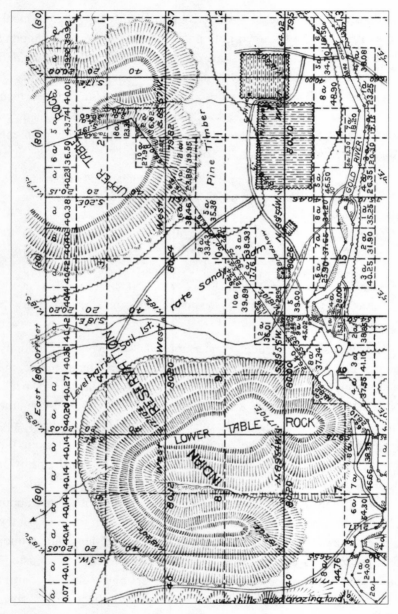

Figure 48. General Land Office Map: Township 36 South, Range 2 West, W. M.;
Table Rocks, Contracts 39 and 47, Butler Ives and George Hyde, 1854. (USDI
Bureau of Land Management, Portland, Oregon.)

marker on the line. Hyde and two of the men circled part way around the rock to find an accessible place and scrambling to the top, hiked back across the mesa to the edge above the point where the rest of the company waited. Peering over the rim, the surveyor calculated the distance "up on top of Rock from Object" at about 146 feet. Surveying the line east approximately 1,000 feet across the mesa from its western edge, Ives and Hyde set the post between miles 22 and 23 of the offset. For three-quarters of a mile the company crossed the rock's flat expanse marked by vernal pools, native bunchgrasses, and blue camas. The surveyors lunged down the eastern face of Table Rock to set a corner post on a ledge part way down the cliff. Climbing the rest of the way down through white oak forest underlain with cat's ears and blue lupine, the men entered the level plain to camp along the Rogue River for the night.[33]

The next morning, May 20, the company ferried across the Rogue and surveyed the offset farther east. On the line's final two miles, they crossed the gently rolling prairie near present Eagle Point, passing the

Figure 49. View of Lower Table Rock from the Rogue River. (Southern Oregon Historical Society, Photo #14386.)

Fryer and Cameron farms to the south of the line and the Mathews claim on the north. They forded Little Butte Creek and, completing the thirtieth (and last) mile of the offset, set a post on the true Willamette Meridian (Figure 1).[34]

Willamette Meridian South, May 20–25, Seventeen Miles

Coming south, the surveyors made good time through open, grassy fields scattered with oaks and volcanic rock. Passing Silas and Edward Day's claim in the first two miles along the meridian line, Ives looked east to see snow-capped Mount McLoughlin, the prominent peak Jesse Applegate had recommended using as a triangulation point. The company came south on May 21 through rolling prairie. Finding little suitable timber to make posts in this country, Addington and Griffiths set the corners with rocks and raised mounds, in one instance marking a section corner with a chunk of petrified wood.[35] Now the party moved into higher country in the foothills of the Cascades. In the 222nd meridian mile, the men reached the foot of a perpendicular, basaltic cliff, present Payne Cliffs in the vicinity of Phoenix, Oregon. Part way up this elevated point, Butler Ives observed another snow-covered peak, Mount Ashland, far off in the Siskiyous as well as Pilot Rock, "a high isolated knob" located in the same general direction.[36] Named later for the town lying about eight miles north of it, Mount Ashland has an elevation of 7,530 feet above sea level, while Pilot Rock, prominent guiding landmark for pioneer travelers between California and Oregon, stands at 5,914 feet.[37] On May 21, the survey party ended one of their best days yet, making nine miles by nightfall.

On May 22, Ives and Hyde descended the hills through Township 38 South and entered the valley of upper Bear Creek, a stream originally named for Army Captain James Stuart, who had died nearby in a confrontation with Indians two years earlier.[38] The main portion of what has historically been termed the Rogue River Valley, the Bear Creek Valley extends from the base of the Siskiyous toward the Rogue, where its major stream joins the river near the Table Rocks. The first few settlers arrived in 1851; Samuel Colver stopped along Bear Creek near present-day Phoenix, Oregon, while Hugh Barron, Patrick Dunn, and Thomas Smith built cabins in the valley's narrow southerly end. When prospectors discovered gold at several locations in southwest Oregon in the winter of 1851–1852, the ensuing rush brought hordes of miners into the region at the same time more settlers arrived. The Anderson brothers claimed land west of Bear Creek near present-day Talent, Oregon, in 1852, while Abel Helman and Robert Hargadine built cabins on Mill Creek at present-day Ashland. In the fall of 1853, eight months prior to the surveyors' arrival in the region, 159 wagons lumbered westward down the Cascade Range's Greensprings slopes, bringing hundreds of new emigrants into southwest Oregon.

Wilamette Meridian South, May 23–24, Six Miles

Crossing the Oregon-California road southeast of present-day Talent on the morning of May 23, Ives and Hyde sent their packers with the loaded mules south along the trail to the foot of the Siskiyous to await their arrival, hoisted their heavy packs, and pushed the meridian into the Siskiyou Mountains.[39] Climbing south along the ridges west of what is now Ashland, Oregon, the company made five miles before camping for the night at Horn Gulch.[40] Threading through thick brush the next day, the surveyors measured mile 234 over high, rock-strewn hills to set the post for the corner of Townships 39 and 40 South, Ranges 1 East and 1 West. Gazing across the Siskiyou's rugged terrain to the south, Butler Ives decided to offset east from the Willamette Meridian (Figure 1).[41]

The company had made just a little over a mile by late afternoon. Four of the men ended the day with a vigorous excursion. George McFall, who had a wonderful time, wrote that night "Went in company with three others to a high peak [Wagner Butte] . . . I had a splendid view from the highest elevation I was ever on, being over five-thousand feet high. We wrote our names on the summit, which is composed of granite some 80 or 90 feet high and partly covered with snow on which we had some rare sport in sliding down."[42]

Willamette Meridian South Offset, May 25, Six Miles

On May 25, the survey party traveled east four and one-half miles along the south boundary of Township 39 South, Range 1 East to cross the West Fork of Ashland Creek. Early that afternoon one of the men killed a young grizzly bear after cornering the animal in a chinquapin and willow thicket.[43] The surveyors' journals record occasional encounters with wildlife. In addition to grizzly bear, Ives and his men would have seen a variety of animal species in the eastern Siskiyous including black-tailed deer, elk, bobcat, cougar, red fox, gray wolf, and black bear. Gazing skyward, the company would certainly have seen ducks, geese, bald eagles, golden eagles, and California condors that inhabited the area.[44] At day's end, the surveyors set the post corner of Townships 39 and 40 South, Ranges 1 and 2 East on a steep granite hillside about three miles north of present-day Siskiyou Pass.[45]

The next day, May 26, Butler Ives and two of the company descended a short distance east to the Oregon-California road where passing miners told of violence in the Siskiyous. The deputies called a halt to the meridian survey. "We did not think it prudent to extend the Meridian Line to the Southern Boundary of the Territory," Ives disclosed, "on account of a late outbreak with the Shasta and Applegate bands of Indians who were committing depredations upon the whites in the Siskiyou Mountains, near where the line will run, but we will endeavor to extend it through by the time we finish running exterior lines."[46] As the surveyor scribbled his notes, a shadow fell across the page. He looked up as an eclipse of the sun silently witnessed the axeman's final blows to the post.[47]

The company hiked two miles north down along the Oregon California road to reach the Mountain House in the southernmost reaches of the Bear Creek Valley. At the simple wood frame hostelry, set amidst broad green fields along Hill Creek, the proprietors ran a lucrative business selling food and accommodations to travelers crossing the Siskiyous. To supply their stores, Hugh Barron and James Russell — partner John Gibbs had died the previous summer during the fighting between white settlers and Indians — raised beef cattle, chickens, corn, cabbage, and onions and stocked hundreds of pounds of flour, sugar, and salt. On the road between the inn and Barron's vegetable garden, the company met their waiting packers.

As southern Oregon warmed in late spring, Ives and Hyde temporarily abandoned the meridian to survey the valleys of Bear Creek and the Rogue. Months of work lay ahead and, as always, speed and efficiency would determine their chances of making a profit. Good weather and their first-rate crew increased their odds for success.

Chapter 10 – 1854

Chaining the Rogue River Valley

"Chain by chain, mile by mile, stake by stake."[1]

On May 27, at dawn, Ives and Hyde broke camp near the Mountain House and, as the sun crested the Cascades, walked north toward Jacksonville miles away on the western edge of Bear Creek Valley. Settlers on the farms they passed resumed work they had abandoned the night before; Isaac Hill split fence rails, Patrick Dunn gouged out an irrigation ditch, while Giles Wells plowed long furrows across his field. At present-day Ashland, the surveyors crossed Abel Helman's claim where, on Mill Creek, the native Ohioan carpenter raised beams for a flour mill not far upstream from his sawmill. Three miles north along the road, laborers notched timbers for another flour mill at Eagle Mills on Bear Creek. Making twenty-one miles in a day, the company set up camp outside Jacksonville.[2]

The next day, the crew spent their day off in town. George McFall "found it quite a stirring little place, it being Sunday the busiest day of the week when the miners come in to lay in their week's provisions and recover their letters."[3] Jacksonville had changed since William Ives' visit two years before. Along California Street, many of the crude log structures and tents had been replaced by wood frame saloons, boarding houses, and stores. The Cram Rogers & Co. Express building stood at the southeast corner of Third and California streets while

Figure 50. Jacksonville, Oregon Territory, 1856. (Kuchel and Dresel, artists; printed by Britton and Rey. Private Collection.)

the Robinson House was on the northeast corner. Adjacent to the express office, carpenters nailed board siding on T. McFadden Patton's Law Office. The Arkansas Livery Stable, Maury and Davis' brick store, Appler and Kenney's mercantile, and Zigler and Martin's blacksmith shop lined the busy main streets. The New England Bowling Saloon — called the Palmetto when William Ives visited it — still welcomed thirsty men (Figure 50).

On Monday, May 29, the company left Jacksonville for their survey location. The deputies divided the men into two crews. Ives chose McFall, Shunk, Cowne, Lewis, and Addington, while Hyde selected Truax, Cannon, Griffiths, Price, and Payne. For the next week, Ives and Hyde measured boundaries for eight townships.[4] The northern tier of three townships that extended east from present-day Rock Point to Eagle Point encompassed the Rogue River and the Table Rocks. The middle tier stretched from the headwaters of Forest and Jackson creeks west of Jacksonville east across the broad plain of what is now Medford, to the foothills of the Cascades. To the south, the townships embraced the valleys of Coleman Creek, Gore

Creek, and Wagner Creek in the vicinity of Talent and Phoenix, Oregon.

As the surveyors chained across the Bear Creek Valley in early June, the wild roses blossomed along the streams. Most days, the sun shone, and with the exception of a few light rains the weather allowed almost steady use of the solar compasses.[5] Ives went into Jacksonville on June 10 to mail the Willamette Meridian field notes to Surveyor General Gardner. "Mr. Hyde is now going to subdividing with his party, while I intend to complete the exterior lines," Ives wrote Gardner. Jacksonville attorney Patton, on his way to Salem to be married, agreed to deliver the precious package. After turning over the materials to Patton, Ives and McFall loaded the mules with fresh provisions the next day and returned to camp.[6]

From June 10 to June 19, George Hyde and his crew worked in the vicinity of Jacksonville. West of town the men crossed the Rich Gulch gold diggings, where miners crowded side-by-side on narrow claims, their shouts and creaking equipment resounding through the ravine. Using a rocker — a wooden box with a screened mesh bottom — the prospectors dumped in gravel, flooded it with water and rocked the box to wash out the gold. Nearby, other miners used sluice boxes — troughs fitted with riffles on the bottom — to catch the precious metal. At one end of the gulch, Chinese miners shared a claim. Working efficiently in groups, and satisfied with lower returns, their thoroughness stood in sharp contrast to the white miners, who skimmed off the easily available gold before moving on. Above Rich Gulch on Jackson Creek, miners worked both sides of the stream for miles. Hyde passed through One Horse Town, or Bolluxburgh, a small mining town of log houses about one-half mile west of Jacksonville. His field notes provide us the only known description of this now-long-abandoned settlement that stood near the road to the Applegate, along what is now Highway 238. The surveyor wrote, "Bolloxburgh is situated near the west Boundry of Sec. 31 & the main street runs E & W about 4 chains from the North Boundry of the same sec. It contains about 30 cabins and is a thriving little place" (Figure 51). Finishing his survey

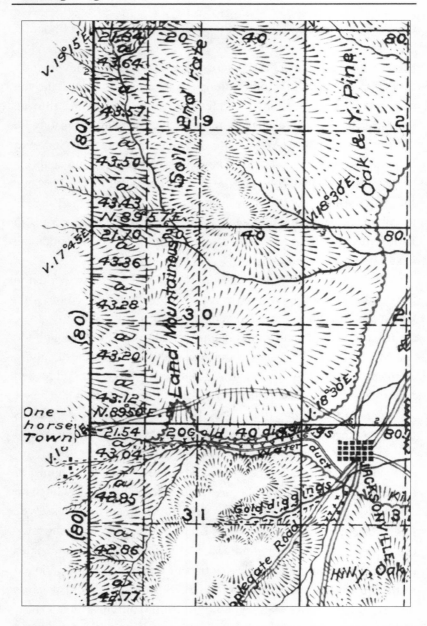

Figure 51. General Land Office Map: Township 37 South, Range 2 West, W. M.; Jacksonville, Oregon vicinity, Contracts 39 and 47, Butler Ives and George Hyde, 1854. (USDI Bureau of Land Management, Portland, Oregon.)

near Jacksonville, Hyde spent the next ten days subdividing the heavily settled 36 square miles of Township 38 South, Range 1 West in the vicinity of present-day Phoenix and Talent.[7]

Willamette Meridian South, June 13 to June 14, Twelve Miles

On June 13, Butler Ives and his men surveyed a short segment of the Willamette Meridian north of its intersection with the eastern offset established on May 20. Chaining north of present-day Eagle Point twelve miles into the foothills of the Cascades, the crew completed the meridian extension the following day. Working now in open country, the men endured long hours in the hot sun, stopping only to take their noon meal in the shade. Over the next three weeks, Ives and his men surveyed the boundaries for eight townships, finishing on July 8, 1854.[8]

Willamette Meridian South, July 9 to July 10, Nine Miles

Ives retraced his steps into the Siskiyous to the north boundary of Township 40 South between Ranges 1 and 2 East, where he had temporarily abandoned the meridian survey during an outbreak of violence between Shastas and miners in late May. All day on July 9, the party chained over granite rock ridges and down into ravines, measuring the meridian south over the Siskiyou divide into the Klamath River drainage. Through fir and pine forest, the axemen cut through brush of laurel, hazel, dogwood, and lilac to make four miles before camping near the summit of the mountains.[9] From a high point the following day, Ives saw "Mount Shasta Colofornia" far to the south. The men finished the last five miles by evening and, notching yellow pine and cottonwood as bearing trees, set the last post on the Oregon-California boundary near Cottonwood Creek. "Finished the meridian," the surveyor noted, "Extended it to the south boundary of the territory and established a post, the first [set] between Oregon and California" (Figure 1).[10]

Ives may not have been aware of another recent official survey in the region. On April 27, 1854, after both California and Oregon claimed votes from the mining camps of Althouse and Sailor's Diggings, the California Legislature appropriated $3,500 for a survey "from the coast to the vicinity of Pilot Knob [Pilot Rock]." Klamath County, California, Surveyor Thaddeus P. Robinson was awarded the job. Robinson started out from a United States Coast Survey monument at Crescent City in Klamath County — formed from a portion of Trinity County in 1851 — and, on June 6, placed the boundary's first monument just north of the mouth of the Winchuck River. Surveying east from that point until June 26, Robinson chained the line about eighty miles east through rugged terrain to the summit of the Siskiyous in the vicinity of Pilot Knob. At that point, since Robinson had passed by the disputed area near the mining camps in the upper Illinois Valley, he halted his survey.[11]

From the last post on the meridian, Ives gazed down on Byron and Rufus Cole's house and barns in the Cottonwood Creek Valley. New York natives, the Coles farmed land not far from the place where Colestine would eventually bear the family name. The deputy noted bearings for Pilot Rock about five miles to the northeast and for the Oregon-California road a short distance to the east. "[The road]," Ives noted, "passes through the lowest place in this range of mountains & is being made into a good waggon road."[12] The surveyor's field notes for the last few miles of the meridian describe the terrain and vegetation in the rugged Siskiyous long before intensive logging and before the time when automobile travelers would glide smoothly through the mountains. "The Northern slope is very broken with steep ravines and ridges," Ives observed. "It is all heavily timbered with fir, redwood [yew or incense cedar], y and S pine [Ponderosa and sugar pine] and underlaid with granite rocks . . . The remainder of the line to the Southern boundary of the Territory, except a very small valley on Cottonwood creek, passes over rolling open hills . . . They rise gradually to the westward into heavily timbered mountains . . . The mountains East of the road are high & rugged. The Southern slopes are generally thinly timbered & portions of them are bald."[13]

∽

Descending the Siskiyous, the company walked thirty miles north to Jacksonville on July 11 and 12, arriving in town at 11 P.M.. On July 13, a blistering hot day, Ives and his men hiked twenty-two more miles north to their survey location near the Rogue River to chain three townships north and west of what is now Eagle Point.[14] Finishing work in the area a week later, Ives returned to Jacksonville for provisions, loading fifty pounds of flour, ten pounds of sugar, and twenty pounds of fresh beef onto the mules. Meanwhile, McFall and the rest of the company hiked fifteen miles down the Rogue to the next survey location in the mountains north what is now the town of Rogue River. On their way back to Jacksonville a week later, the company stopped over at McFadden's ranch where they helped the judge and earned extra money at the same time. In George McFall's words, "All hands lent a hand to McFadden in harvesting which we found to be very different work from surveying, it blistering our hands and give us a back ache and a strong desire to be back once more in the woods again." Tired and sore, the company returned to Jacksonville and camped outside town.[15]

During the first half of July, as Ives and his men worked in northern Jackson County, George Hyde surveyed north and east of present-day Medford, an area with good grazing land in the hills and fine soil in the creek bottoms. "There is from 30 to 35 claims taken in Township [36 South]" Hyde noted, "but a considerable number of the claimants are Bachelors & are absent in the Mines."[16] The surveyor's comment about the missing "Bachelors" records the seasonal temptation that lured settlers from their farms to the gold mines. A single man could make a show of settlement to his neighbors by building a small cabin and plowing an area of ground before disappearing for months at a time in the gullies west of Jacksonville. A married farmer might leave his wife and older children in charge of the claim to try his hand at mining.

Finishing up section lines north of present-day Eagle Point during the last two weeks of July, Hyde joined Ives at the end of the month in Jacksonville. From there, he wrote Surveyor General that he and Ives were out of money and would soon need funds to pay the hands, adding, "In fact we have been out of money some time, are buying our provisions on credit."[17]

⌐

John W. Davis resigned as Territorial Governor on August 1, 1854. Although he was a Democrat, Davis had survived only nine months of what one historian described as the "attack and invective in the best of Oregon's uninhibited newspaper style." As the editor of the Whig *Oregonian* observed, local Democrats would not accept Davis because "he had neither driven his team across the plains nor been to the mines."[18] On the same day Davis resigned, Charles Gardner moved from Oregon City to Salem, the final step in the long-expected reloca-tion of the surveyor general's office. "The maps, charts and notifica-tions, with the proofs and records of the 4500 claims in this territory," Gardner wrote the General Land Office in Washington, DC, "were safely secured in boxes lined with tin and made water-tight by sodering and brought by boat to Champoeg and then by wagon to Salem, along with cabinets and furniture." Large items, including the draftsmen's tables and desks, heavy cases, and the cumbersome iron safe had been stored temporarily at Canemah to await high water and transportation to Salem by steamer the following November.[19] Gardner assured the commissioner, "All the precautions you have enjoined, relative to the safety of the office and its archives, have been anticipated by me. My present office building I consider much safer than the one we left: and the excellent messenger I have in service, for the care of it all, is unsurpassed in vigilance."[20]

As Gardner settled in at Salem, Butler Ives wrote him from Jack-sonville on August 2. Advising the surveyor general that they had recorded exterior lines for "all of Gold [Rogue] River valley as far down as Range 5 West that will likely need surveying for many years to come," Ives listed the townships that he and Hyde had decided to subdivide in full.[21] In closing, Ives listed areas needing future survey: the

Applegate River Valley, the small valleys of Wolf Creek, Grave Creek, and Louse Creek, and the valley of the Rogue River west of present-day Grants Pass.

⤷

Ives and Hyde spent hot weeks in August dragging equipment over rough terrain. The two deputy surveyors worked separately, George Hyde subdividing a township southeast of present-day Eagle Point while Butler Ives worked north of Ashland.[22] In mid-August, Ives and his men moved north of the Rogue River to subdivide Township 36 South, Range 2 West which contained about twenty donation land claims as well as the Fort Lane military reserve.[23] In this challenging township Ives had his turn chaining the line across Upper Table Rock. Then, while measuring north between Sections 16 and 17, the deputy surveyed the line over the Rogue River and into a slough before scrambling up the steep south slope of Lower Table Rock.[24] "The summits of Upper & Lower Table Rocks," Ives described as "level stoney table land, elevated about 840 feet above the bottom land on Gold River & . . . terminated by abrupt ledges of a scoriaceous lava from 50 to 250 feet high, which is underlaid by a thick formation of sandstone."[25] Finishing up another township near the end of August, Ives returned to Jacksonville where he and George Hyde, along with their crews, swore to completing Contract 39 (Figure 52).

On August 30, the deputies looked over Contract Number 47, recently mailed them by Surveyor General Gardner. An extension of Contract 39, the new document directed the subdivision of the ten final townships, an estimated 300 miles of survey, for which they had already surveyed the exterior boundaries.[26] Initially relieved to find the new contract waiting for them, Ives and Hyde were then stunned to read that they would have to take a cut in pay. Three months earlier, General Land Office Commissioner Wilson complained to Gardner about the "high rates" received by Oregon's deputy surveyors, suggesting that the Territory's present lower cost of living warranted reducing

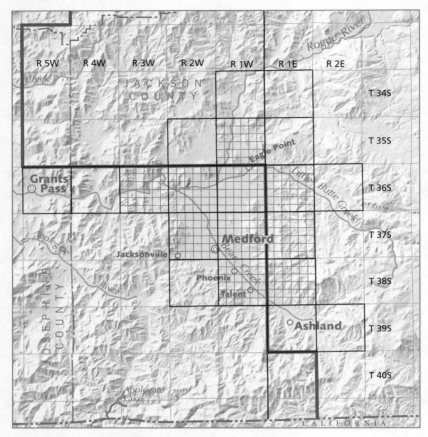

Figure 52. Survey Contract 39: Townships and subdivisions surveyed in Jackson County, Butler Ives and George Hyde, 1854. (Map image from Stuart Allan, et al., *Atlas of Oregon*. Eugene, Oregon: University of Oregon Press, 2001.)

salaries. Gardner defended the surveyors' pay rates, explaining that "Contract 39 . . . [was] made with the allowance of $17 per mile for the exterior lines of Townships, and $11 per mile for subdivision lines, making when *the whole are run*, the average of $12 per mile — because I was advised by my able assistants that these rates were allowed in many previous contracts, to which no exception was taken at your Department."[27] Reducing his deputies' pay was unfair despite a temporary lower cost of living, Gardner explained, noting that "the difficulties of accomplishing the surveys of the remaining townships, and remaining Standard Parallels, have increased in much greater proportions." Gardner reinforced his argument, "The contract which I made with G. C. and C. T. Gardner, in Washington Territory, is not expected to pay their expenditures and was refused by Hunt. In the contract of Latshaw, he was broken up before he finished his subdivisions."[28]

At Wilson's insistence, Gardner cut the rates for Contract 47 to $12 per mile and in one last devastating action, retroactively reduced the deputies' pay for the exterior and subdivision lines on Contract 39 to the uniform rate of $12 per mile. Completing that contract in good faith, in debt to their men and to storekeepers for provisions bought on credit, Ives and Hyde were forced to accept $12,687.00 rather than the $13,467.00 they had anticipated.[29] Far from Salem, unable to protest, and obligated to keep working, they signed the new contract, "We, Butler Ives and George W. Hyde do hereby consent that the compensation named in said contract for running and marking Standard parallel, Township and Subdivisional lines be Twelve dollars per mile and hereby relinquish our right under said contract to any compensation . . . exceeding Twelve Dollars per mile" (Figure 53).[30]

Fall, 1854

Early in September, Ives and Hyde surveyed along the Rogue River west of present-day Gold Hill. The angle of the sun lowered and, in the afternoons, clear, golden light washed over the hills. September 9, Ives and his crew hiked east several miles to McFadden's ranch where they surveyed the jurist's land claim. Moving southward along the main trail, Ives next surveyed the heavily settled upper Bear Creek Valley.[31] Both Abel Helman's flour mill at Ashland Mills and that of the Gilead Mill Company at Eagle Mills were now in operation. In the former settlement, carpenters Eber and Jacob Emery raised beams for the new boarding house in front of the flour mill, while farther up the Bear Creek Valley farmers toiled in the endless rhythm

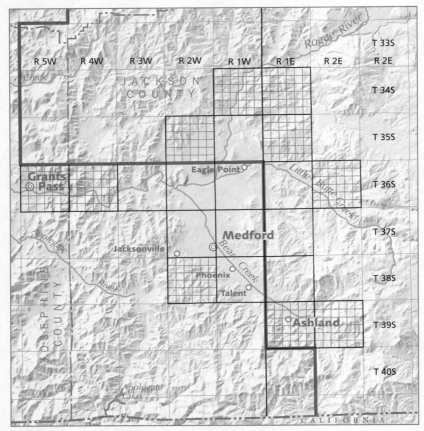

Figure 53. Survey Contract 47: Subdivisions surveyed in Jackson County, Butler Ives and George Hyde, 1854. (Map image from Stuart Allan, et al., *Atlas of Oregon*. Eugene, Oregon: University of Oregon Press, 2001.)

of breaking prairie and digging ditches. Back from the muddy trails that led between claims, they pulled oak stumps from the earth and in the low foothills cut trees for buildings, fences, and firewood. As one settler observed, "A few years sufficed to cut down all the saw timber and the once open forest soon became a forest of young pine and other trees, with a mass of rotten treetops and limbs, the refuse of the waste of logging when only the straight limbed bodies of the trees were used."[32]

As they improved their claims, these hard-working, land-hungry Euro-American settlers changed the appearance of the Bear Creek Valley and, in the process, shattered centuries of Indian culture. For thousands of years, native peoples preserved food sources in the valley and foothills with late summer and fall burning that stimulated the growth of native grasses and kept oak woodlands clear of encroaching conifers. When Ives and Hyde arrived in southern Oregon, the profound changes introduced by emigrants had already devastated these ancient traditions. During ensuing generations, the vast donation claims, some a square mile of fields, forest, and riparian lands, would give way to housing tracts and road systems. Swathed in old-growth conifers, the north slopes of the Siskiyous would, after years of logging and intense fires, for a time yield only eroding soils and charred stumps. The smoke that curled from the chimneys of scattered farmhouses along Wagner Creek would drift away, only to reappear a century later as emissions from thousands of wood-burning stoves and automobiles. In 1854, however, the farmers' gains in the fields along Bear Creek promised only progress.

On September 26, Ives and his men returned to Jacksonville for provisions and the next day traveled fifteen miles to Township 34 South, Range 1 West where the Rogue rushes out of the volcanic mountains near present-day Shady Cove. The country, Butler Ives observed, was unsettled. "East of the River the hills rise abruptly from five to ten hundred feet high. There is a narrow bottom along Gold River of first rate farming land." The company finished work on October 11 and moved west of the Rogue River to survey a township in the vicinity of Sams Valley, northeast of Gold Hill, Oregon.[33] In the meantime, Hyde and his men surveyed section lines in Township 38 South, Range 2 West south of Jacksonville, where he found "the new town of Sterling . . . already quite a thriving mining town, the miners expecting to make their piles during the wet season."[34]

October frosts now skimmed camps at dawn, and the mules breathed steam in the cold air. The oaks turned to gold and, by the time the surveyors stopped work in the evening, the western light cast long shadows across the dark, wooded streams. Energized by the cooler weather and the approaching end of their contract, the deputies made good time. From October 7 to October 18, Hyde subdivided along Big Butte Creek while Ives finished up in the hills north of Sams Valley. His handwriting scrawling as he reached the end, Ives set the last section post west of Chimney Rock Butte and made his final notation — "Finished Oct 19th 12 ½ oclock pm AD 1854."[35] Butler Ives and his men returned to Jacksonville, where they found Hyde and his crew waiting for them. On October 21, all hands signed their oaths for satisfactory completion of Contract 47 and then celebrated. The next morning, the company separated. George McFall bid his colleagues goodbye and started for San Francisco and his return east.[36] Ives and Hyde headed north to Salem to report to Surveyor General Gardner.

⌒

    Gardner issued eleven survey contracts in 1854. In addition to those awarded Ives and Hyde, he assigned contracts to his own sons, Charles and George Gardner and to Harvey Gordon, Josiah Preston, Lafayette Cartee, Daniel Murphy, Matthew O. C. Murphy, Nathaniel Ford, Ambrose Armstrong, Andrew Patterson, and Joseph and John Trutch.[37] Gordon and Preston spent the spring surveying boundary lines west of Corvallis — its name was changed from Marysville in 1853 — in the foothills of the Coast Range and south of Eugene. Cartee, who had missed out on Contract 39, worked in five townships east of Oregon City. Murphy took some of the townships that were assigned to Joseph Hunt at the time of his death — areas located in the Coburg Hills and the McKenzie River Valley. Gardner gave Ford Contract 44 for work in the Umpqua country. Gardner's sons took a joint contract in Washington Territory to survey exterior boundaries and subdivision lines.[38] Armstrong ran lines in the small valleys and foothills west of Corvallis and McMinnville, while Patterson surveyed six townships southeast of Eugene. Gardner awarded the last contract of the year to Joseph and John Trutch. Betrothed now to Julia Hyde, Joseph Trutch worked with his brother northwest of Portland. Gardner also awarded thirty-two contracts for donation land claim surveys during the year in townships from Salem south to Roseburg. Walter Howard, J. Addison Pownall, David Stump, and Abraham Sulger, who were new to the claims surveys this year, joined experienced survey-ors Samuel Snowden, Matthew O. C. Murphy, and Andrew Patterson.[39]

His stomach clenching, the surveyor general fielded attacks from the General Land Office where officials, far from Oregon Territory and out of touch with the realities of administering the public land surveys, generously handed out hard-nosed rulings while stingily hold-ing back funding. In one particularly odious regulation approved by the Secretary of the Interior, Commissioner Wilson forbade Gardner to continue his practice of making partial payment to the deputy sur-veyors on contracts already issued as well as future agreements. To help the cash-strapped deputy surveyors, Gardner had made it a prac-tice to grant them partial payment after examining notes of groups of townships for correctness. Now Wilson delayed payment for months by prohibiting any recompense prior to absolute completion of their contracts, submission of the notes to Washington, and a notice of ap-proval from the General Land Office. "Any advances by an officer of the Government I consider as prohibited by law," Gardner wrote Wil-son, "but after the evidence of work done (for instance, the entire survey of the exterior lines of a township,) I have considered myself authorized, to enable him to pay for his supplies, and his hands, to make him a payment in part." Predicting that the ruling would make it more difficult to hire qualified surveyors, Gardner advised Wilson, "The effect of the order upon the future surveys in these Territories, where the lowest rate of interest is 3 percent a month, you will, as readily anticipate, as from any suggestions of mine."[40]

From the Rogue River Valley where, despite the recently com-pleted public land surveys, settlers still awaited official designation of donation land claims, anxious correspondents peppered Gardner with questions about the status of their lands. "John Coakley built this house in May 1852," Alexander Gordon wrote from his home near the Oregon-California road east of Bear Creek, "and perhaps he had a survey made but there had never been any fire built in the house. He was then absent til about the 15 of September [1852] when he come and had a survey of 320 acres including my 160 acres . . . Sometime after this survey he went into the vacant house and claimed to hold the 320 acres, one half on account of a wife which he says he has in the atlantic states tho she has never bin here I don't know as he has any . . . If he can hold, I must lose my two years cultivation and find a new farm. If you are willing to give me your opinion on this amount and will do so, you will much oblige."[41]

Amanda Hardin, an illiterate sixteen-year-old widow living up Kane Creek southeast of Gold Hill, posed her question in a letter written for her by Jackson County attorney W. G. T'Vault: "On the 26th day of June 1853 said John R. Hardin and myself were married and on the next day we moved and went to live on the Claim with the said Patrick . . . until the Indian war commenced . . . we left the claim about the first week in August and on the 11th day Hardin was shot by the Indians . . . and on the 13th died."[42] Explaining that someone else now lived on the portion of the claim once occupied by Hardin and Patrick, the young widow added, "I am anxious to know whether I as the widow of said Hardin having since his death had male Issue, have any right or interest in said claim." Gardner answered T'Vault, "The widow Amanda E. Hardin can hold the qr. Sec for herself and child on proving up to the day of her husband's death and take another quarter section in her own right."[43]

Despite his chief clerk, assistant clerks, three draftsmen, and two "copyists," the surveyor general fell behind with the paperwork. "The pressure upon my office has been extraordinary," he wrote the Commissioner early in October, "26 notifications have been filed . . . and proved before me in one day." Gardner explained that it would be "utterly impossible to have my Annual report prepared in compliance with your instructions of the 1st of August, in time for the meeting of Congress." Should this estimate miss the next Steamer from Portland to San Francisco, Gardner advised the Commissioner, "it may not reach you before December."[44] The surveyor general finished his annual report to the General Land Office on November 21, 1854. He inserted a colored lithographed map to illustrate the townships surveyed in the Rogue River Valley, although unfortunate mistakes on the document placed Jacksonville four miles east of the Willamette Meridian instead of ten miles west of it, and sited Upper Table Rock fourteen miles east of its true location.[45]

Deputies had surveyed 429,493 acres in Oregon Territory from 1852 through 1854. Processing thousands of notification documents and mapping for both the rectangular surveys and donation land claims contracts overwhelmed clerks in the surveyor general's office. "The staff had no means for selling lands authorized in 1853 by Congress," historian Stephen Dow Beckham has observed. "It had to record all of the Donation Land Claims and note the connections of the plats of those claims with the survey lines in the townships before it could proceed with the orderly sale of public lands."[46] To expedite settlers' acquisition of land, Congress approved the Act of July 17, 1854 (10 Statute 305), which reduced the residency requirement for purchase of a claim from two years to one and granted appropriations for surveys in Washington Territory. The act also established three separate land offices headed by staff to record the land entries and to handle money collected in payment for land claims.[47]

In November came official word that the US Land Office would open on December 11, 1854, at Oregon City with Ralph Wilcox in charge.[48] Wilcox had been temporary clerk of the Supreme Court the previous year. "The Surveyor General of Oregon will cease to receive 'Notifications' of donation claims, from and after the first day of December next," stated the *Oregon Spectator,* after which time — on and after the 11th of December — "all notices of Donation claims, under the several land laws, all applications of land claims as well as for Pre-emption rights, (etc.) . . . have to be made direct to the Register and Receiver of the Oregon land district at Oregon City."[49]

Ignoring Gardner's ever-increasing responsibilities, the General Land Office acidly criticized him for exceeding his allotment of clerks and draftsmen. The surveyor general defended himself by citing the extra work of transferring records to the new land office in Oregon City as well as his time spent helping open the new surveyor general's office in Washington Territory. Three clerks who preceded Surveyor General James Tilton to the West Coast had arrived at Gardner's office in the fall of 1854 to ask his assistance.[50] He had been extremely busy, Gardner told the Commissioner, and "consequently, although the apportionment was so far reduced I could not discharge any of my office force until the end of the year . . . this I did, by discharging five of the lowest Clerks (having given them due notice)." [51]

Denouncing Gardner's cost-saving move as a political tactic, Asahel Bush charged that Gardner had singled out the Democrat clerks in the surveyor general's office by giving them notice that their jobs would end on January 1, 1855. "The transfer of the notification business," the relentless Bush alleged, "makes it necessary for him to dispense with a portion of his clerks, and the few democrats, occupying the inferior places, are made the victims."[52] Concerning the clerks Gardner had retained in service, John Trutch, George Belden, and Wells Lake, Bush asserted, "Everybody here knows . . . that they have not one democratic sympathy and never cast one democratic vote; that they have always been whigs, and that if they are anything else now, they have made the discovery very *recently*." Calling Gardner incompetent, Bush accused him of favoring Whig John Preston by referring donation land claimants to Preston's private office and charged the office clerks with giving the former surveyor general maps and abstracts of claim notifications for which conflicts required the settler to seek the services of an attorney. "Every reader," Bush asserted, "will see the great advantage this favoritism gives Mr. Preston over all the lawyers in the Territory."[53]

⟝

Butler Ives and George Hyde sent the last of their field notes for Contract 47 to Gardner in Salem in early November. Bitter about his loss of pay, Ives asked Hyde to collect money due him for Contract 47. He sold his horse, tent, and most of his surveying equipment and paid his bill at the Main Street House. Three and one-half years after he had first walked into town with his brother, Butler Ives left Oregon City for San Francisco and the trip home to Detroit. Of the surveyors who had come to Oregon in the spring of 1851, William Ives, James Freeman, and Butler Ives were gone; Joseph Hunt was dead; George Hyde alone remained, and he had some decisions to make.

Chapter 11 – 1855

Back to America

"Nearly all the early . . . deputies have left the coast."

The apparent flood of opportunities for surveyors that, in 1851, drew John Preston, his brother-in-law George Hyde, Joseph Hunt, William and Butler Ives, and James Freeman to Oregon Territory had, after a few short years, subsided. With the exception of Hunt, who died, the surveyors had departed one by one. Surveying had been an exciting and potentially lucrative business in 1851, but by 1855, fewer contracts were available, salaries had been reduced, and relentless political pressure seriously limited Surveyor General Gardner's discretionary power. Local critics and micromanaging bureaucrats in Washington combined to generate conflict and disillusionment. Four years after they had arrived, the surveyors' golden time in the Oregon Territory ended, dissipated in clouds of discord and disappointment.

Winter, 1855

On January 8, 1855, Julia Hyde and Joseph Trutch married at St. Paul's Episcopal Church in Oregon City. After the ceremony, John and Lucy Preston held the wedding luncheon at their home.[1] This would be one of the last social events given by the Prestons. For both personal and professional reasons, the couple had decided to abandon

Oregon for Illinois where life would be easier, safer, and more profit-able. Preparing for this move, Preston had, in December, 1854, agreed with his partner James O'Neill to dissolve their company. Hereafter O'Neill would run the mercantile business single-handedly.[2] On another front, the new US Land Office at Oregon City cut into profits in the former surveyor general's private land law business. On top of Preston's financial frustrations came chronic public criticism from the General Land Office. Eager to place blame for delays in set-tlers receiving official title to their land claims, Commissioner Wilson publicly accused Preston of sabotaging the process while in office by tardily submitting township plats and claims maps to Washington.[3]

Although political conflicts and financial problems would prob-ably have sufficed to send Preston back to Illinois, undoubtedly the most compelling reason was personal. Lucy Preston had become preg-nant, and, at the age of thirty-nine, she was anxious to return to the comfort of the States.

Preston, who had been elected Clackamas County Commissioner the previous June, assured his constituents that he would retain his elective seat until his return to the Territory. Local supporters who had hoped that Preston might represent them in Washington were dis-appointed.[4] Their dismay at his impending departure was noted by the editor of the *Oregon Spectator*: "This announcement has thrown all the whigs back . . . for the reason that [they] and a large majority of the democrats had decided he was a fit and proper person to represent Oregon in Congress."[5]

In one last entrepreneurial venture, Preston devised a new project with his brothers-in-law, Trutch and Hyde, to compile a map of Or-egon and Washington territories. The team was, Preston advertised in regional newspapers, "desirous of obtaining all possible information as to the location of Mills, Postoffices, etc., together with any sketches of the country outside of the Government survey that may be deemed reliable."[6] Joseph Trutch resigned his clerk's position in the Surveyor General's Office to work on the map. With firsthand experience in the country, access to the public land survey plat maps, and skill at drawing, Hyde and Trutch accumulated information for the first detailed commercial map of the Oregon Country.

By compiling the geographical details that his surveyors had recorded in their field notes and that office draftsmen had translated into township maps, Surveyor General Preston created a lucrative possibility for himself, that of selling this information to a commercial mapmaker upon his return to "the States." The General Land Office maps, seen as current and accurate, would become valuable material in the hands of skilled mapmakers, who could then distribute "cartographic information to a swiftly expanding population."[7]

Preston gave George Hyde his power of attorney to handle his business affairs "during my absence in the states," and asked James K. Kelly, a graduate of Princeton University and Carlisle Law School, as well as a prominent attorney and political figure, to attend to residual land law matters. Having arrived in Oregon the same year that Preston came, Kelly was widely involved in territorial issues and a capable representative to act on Preston's behalf.

In mid-February, John and Lucy Preston boarded the steamer at the Oregon City dock and waved farewell to the crowd of well-wishers, among them the newly wed Julia and Joseph Trutch and John Preston's brother-in-law, George Hyde.[8]

～

From Salem, Surveyor General Gardner issued contracts and supervised the clerks and draftsmen's work on the public land survey, relieved that responsibility for processing paperwork for donation land claims lay now in the hands of the land office clerks. In February, with the number of land claimants expanding — 6,000 emigrants arrived in the territory in 1854 — the General Land Office suddenly reversed its decision to center the notifications process in Oregon City and directed Gardner to accept donation land claim applications in Salem. Meanwhile, the territorial land office continued to process them by the hundreds in Oregon City.[9]

Gardner awarded ten contracts in 1855, all between January and July, the only months in which he had sufficient funding to support the surveys. Standardization among survey procedures was apparently a goal of bureaucrats in Washington. The General Land Office sent Gardner new contract forms identical to the forms being used in other territories. Indeed, Gardner, acknowledging the receipt of the forms, wrote the commissioner that in the "six rolls of the new form of Survey contracts the name of Oregon can be inserted for those of 'Minnesota & Iowa' in the print."[10] Recipients of contracts also received up-to-date guidebooks, the brand new *Manual of Surveying Instructions* published February 22, 1855, and the new version of William Burt's *Key to the Solar Compass and Surveyor's Companion*. Burt assured Gardner that his new "convenient pocket companion" would be a useful guide for surveyors.[11]

For contracts in 1855, Gardner signed Zenas Moody, Harvey Gordon, Ambrose Armstrong, Addison Flint, Dennis Hathorn, Nathaniel Ford, Lafayette Cartee, Wells Lake, and George Hyde. Because much of the central Willamette Valley had already been surveyed into townships and sections, these new contracts took the deputies far away from populated areas. Moody, for example, ran lines in the Cascade foothills between the North and South Forks of the Santiam River, while, with other contracts, Armstrong first worked in the coastal mountains west of Willamina and later chained lines west of the Umpqua River near Saddlebag Mountain. Gordon worked near Drain and Yoncalla, Flint surveyed east of Winchester and Roseburg on the North Umpqua, Ford took a contract in the northeastern reaches of Jackson County, and Cartee worked east of Portland in the vicinity of the Sandy and Clackamas rivers.

Gardner awarded Contract 54 to Wells Lake and George Hyde on February 19. This important assignment included areas west of the Oregon-California road in Douglas County and along the Rogue River — townships that Ives and Hyde had recommended for survey, as well as the isolated, gold-rich Illinois Valley in southwestern Jackson County, an area now included in Josephine County (Figure 54).[12] Wells

Lake, who resigned as chief clerk in the surveyor general's office, took the lead on this project — his sole contract in Oregon. Hyde's familiarity with the region made him a logical partner. Gardner assigned the two deputies the boundaries and subdivision lines of ten townships north and west of the Rogue River in what is now northern Josephine County, as well as ten townships in the Illinois Valley. Acknowledging his lack of information concerning the region, Gardner further directed the deputies, "As your surveying operations are to extend to a part of Oregon of which there is but little information as to its correct geographical location and extent of country fit for settlement, you are expected . . . to furnish this office with all useful information that may come to your knowledge."[13]

Wells Lake went ahead to assist settlers in the Rogue River Valley with their notification documents. "In consequence of delays by bad weather and worse roads," he wrote the surveyor general from Jacksonville on March 26, "I did not arrive here until the 18th inst. between which and the 1st inst. the settlers had nearly all made their notifications. I was expected on the 1st and had I have been in time should have done nearly all the business." Lake assured Gardner he would soon be in the field, adding, "The weather has been very fine here since my arrival in the valley."[14] On April 1 at Jacksonville, devout Episcopalian Wells Lake vowed to execute Contract 54 "well and faithfully as I shall answer to God at the great day."[15] The next day, Lake, along with Thomas Payne, James Dickey, Joseph Addington, and Sewall Truax, started work. On April 21, more surveyors pledged their oaths at Oregon City: Josiah Case, Edward R. Shunk, and John D. Griffiths as chainman, and John J. McConnell, William W. Bixley, and Thomas J. Small as axemen. Surveying south from Wolf Creek to Jumpoff Joe Creek and the Rogue River, Lake and his men had been in the field a month when George Hyde arrived. Just a year after beginning his southern Oregon expedition with Butler Ives, George Hyde arrived in southern Oregon the first week in May. He worked in the vicinity of present-day Wolf Creek, while Wells Lake and his crew surveyed nearby.

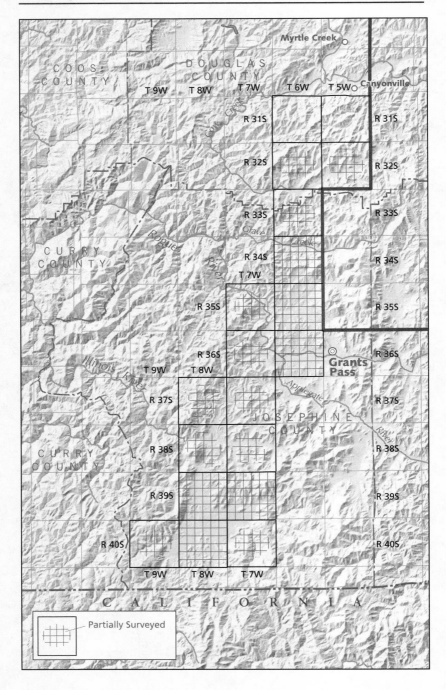

Relations between Native residents and whites were becoming increasingly volatile and dangerous and Lake and Hyde, with their small survey parties in the Rogue country, feared they might be attacked. Certainly the dispossessed bands of Indians were desperate. They had suffered terribly over the winter months from severe cold and hunger and faced starvation if they could not leave the Table Rock Reserve to find food.[16] Finally, tensions exploded between miners and Indians on the Klamath River in northern California. After a miner on the Klamath was killed, volunteers murdered four native men and women near Kerbyville in southwest Oregon's Illinois Valley. Later, when Indians killed twelve miners along the Klamath, Scott, and Shasta rivers, soldiers captured those believed guilty of the killings and took them to Fort Lane, where the volunteer militia demanded that the Army turn the prisoners over to them.[17]

On June 19, Wells Lake and his men finished subdividing Township 35 South Range 7 West between Galice and present-day Merlin, Oregon. A haze of nervousness drifted over the country. "In consequence of the late Indian troubles," Lake wrote Gardner from camp on the Rogue, "we have been delayed in extending our surveys into Illinois Valley. The difficulty has been settled and the indians taken on the reserve by Capt. Smith & Dr. Ambrose the agent. We start over the mountains between R. River & Illinois Valley tomorrow morning During the Indian war, four white men and 5 or 6 indians were killed. The indians were ranging in the Mountains over which we are now to run."[18]

Crossing the Rogue at Vannoy's Ferry about a mile above the mouth of the Applegate River, Lake and Hyde moved south toward the Oregon-California border. Through July and early August, they chained through the Illinois Valley where, by 1855, over fifty

Figure 54 (at left). Survey Contract 54: Townships and subdivisions surveyed in Josephine County, Wells Lake and George Hyde, 1855. (Map image from Stuart Allan, et al., *Atlas of Oregon*. Eugene, Oregon: University of Oregon Press, 2001.)

families — Turners, Newmans, and Northcutts among them — had taken claims on the fertile plain. Lake and his crew plotted boundaries while Hyde and his men subdivided the townships into sections.[19] Outside the cultivated fields, the surveyors climbed higher into white and black oak, madrone, and Douglas fir forest. On south-facing slopes, they clawed their way through buckbrush, manzanita, and abundant poison oak that underlay the Ponderosa pine and cedar. Depending on the demands of their work, the crews shifted at times between the two deputy surveyors. Shunk, Griffiths, and Small usually assisted Hyde, while Case, McConnell, Truax, Addington, and Bixley helped Wells Lake.

The surveyors encountered miners of the Illinois River's tributary streams. In the mining district known as Sailor's Diggings, men crowded on claims along Sailor, Scotch, and Allen gulches, while in other areas the placers had already been exhausted. On the East Fork Illinois River in Township 40 South, Range 7 West, the surveyors found the formerly gold-rich stream almost mined out.[20] In the upper Illinois Valley, the surveyors visited Waldo, supply town for the mining region. Developed in 1853 as a mining camp of temporary canvas tents and crude log structures, Waldo was named for a California gubernatorial candidate when the site was mistakenly assumed to be in that state rather than in Oregon. By the time Lake and Hyde visited the community, Waldo had a bowling alley, butcher shop, blacksmith, brewery, stores, hotels, and saloons along its single east-west street. Pack trains running from Crescent City to Jacksonville were linked to the town by a branch trail, carrying equipment and supplies to the mines (Figure 55).[21]

As the hot summer wore on, mining dried up along the creeks and tensions simmered between Indians and restless miners. Hearing reports of random killings in the mountains, Lake and Hyde hurried their subdivisions and, at the end of the first week in August, finished the contract.

The surveyors separated and headed in different directions; Wells Lake, James Dickey, and Thomas Payne went to Jacksonville to swear their oaths to the district court clerk while Truax and Addington traveled

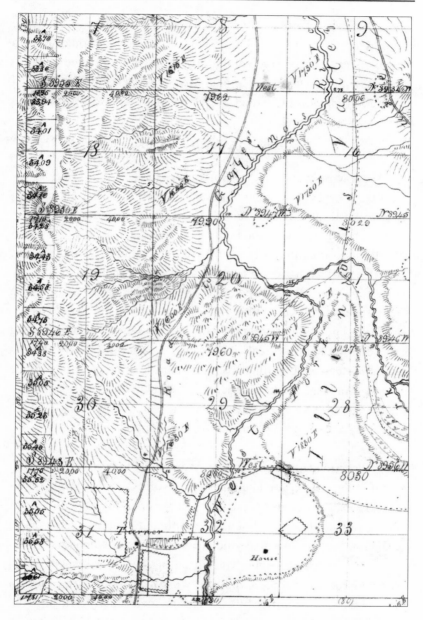

Figure 55. General Land Office Map: Township 39 South, Range 8 West, W. M.; Wells Lake and George Hyde, Contract 54, Illinois Valley, 1855. (USDI Bureau of Land Management, Portland, Oregon.)

to Grave Creek, where McDonough Harkness, a justice of the peace and now proprietor of Bates and Twogood's old inn, would witness their pledges. George Hyde and a few of the men traveled north to Salem where, on August 28, Shunk, McConnell, Griffiths, and Small swore to completing Contract 54. Hyde made his own affidavit in Salem on September 7.[22]

Fall, 1855

The township and subdivision surveys in the Rogue River Valley complete, Surveyor General Gardner planned to hire a claims surveyor to head the donation land claim surveys in southern Oregon. He would issue ten donation land claim contracts in 1855.[23] Settlers backed men for claims contractor who they believed would support them if conflicts arose concerning their claim surveys. Robert B. Metcalf and Sewall Truax competed for this important assignment. Metcalf was a well known Rogue Valley resident, and Truax had assisted Ives and Hyde there in 1854. The competition centered along both personal and political lines. Several Jackson County settlers backed Metcalf and, hailing from Jacksonville, the candidate made his own strong bid for the job, declaring, "There has been an appointment of Sub Agent in the place of Wm Martin resigned offered me but I will not accept it if you can give me this contract."[24]

In recommending Truax to Charles Gardner the previous fall, Preston described him as a "competent and reliable Surveyor, a gentleman of integrity and good habits [with] a solar compass and . . . considerable experience in the uses of that superior instrument." Wells Lake endorsed Truax and sent the applicant's signed petitions to Gardner at Salem.[25] An anonymous correspondent, who signed himself as a "Citazen of Rogue River," did his best to sabotage Truax, writing Gardner "of some fraudulent work that has been imposed on the General Survairs office by false reports made out by Sewall Truax one of the Cumpas men imploid by Mr.s Lake and Hide." Truax, "Citizen" reported, had told him that "when he was in the employment of Mr. Lake and Hide that one of the contractors . . . sent him

and the hands one Sundy morning to meander one side of Rogue river."
Citazen gossiped further:

> Truax told me that when they went to the spot where they was
> to commence thare work that they commenced and set two stakes
> and then he said that they went in to the bushes and maid out his
> notes for the balance of his days work getting into camp earlyer
> than expected he was ast by either Mr. lake or Hide whitch ever
> one was in camp how it came that they got in so soon he said that
> he told him that they had no troubble and got along mutch sooner
> than they expected.[26]

Despite "Citazen's" eager attack on "sutch swindlers of the
publick fundes," Gardner awarded Truax the contract.[27]

As the year drew to a close, pressures on Charles Gardner heightened as funding for the surveys grew tight and as both local and national politicians hounded him. While the true nature of Gardner's effectiveness in his position is obscured by layers of political vitriol, it is certain that the barrage of invective leveled at the office from bureaucrats in Washington, DC, and by Asahel Bush in Oregon undermined his leadership and demoralized his surveyors.

From Washington, an outraged Thomas Hendricks, who had replaced Wilson as Commissioner of the General Land Office, reprimanded Gardner for hiring extra staff. "This Procedure on your part is quite a matter of surprise it being a sudden and unexpected increase in the number of clerks and draughtsmen, doubling it, without reporting to this Office the *necessity* which impelled you to that course, or without even consulting us. You are requested to report upon this subject at the earliest period."[28] Gardner justified his need for eight clerks, "The work performed in this Office since the end of the last fiscal year . . . and the work transmitted since that date to your office, and to the two Land Offices, could not certainly have been accomplished with the aid of four Clerks only; and as much more will be required, in connection with the Surveys of the Territory, for some years to come." Humiliated and angry, Gardner added, "If the censure conveyed in your letter . . . has been communicated to the Hon. Secretary of the Interior, I request, in justice to me, that you will also communicate to him this explanation."[29]

The *Statesman* editor pounded Gardner for incompetence, for being almost two years behind in work and for alleged disloyalty to the Democratic Party. During the legislative assembly's winter session, Democratic members forwarded a memorial to President Pierce, asking him to remove the surveyor general from office.[30] "Under the administration of Millard Fillmore" the authors testified, "Oregon had a surveyor general who was young, vigorous and competent, and an active political partisan, but against whom the democracy of the Territory preferred the serious charge of extortion in office, in demanding fees for services in the ordinary discharge of his duty." Although the Democrats had initially welcomed Gardner, they now called his appointment "exceedingly unfortunate." While admitting that he had not charged illegal fees as Preston was accused of doing, the memorialists charged that Gardner had refused "to do the proper duties of his office, and persisted in so refusing until on complaint of citizens here, the commissioner of the general land-office specially instructed him in this behalf."[31]

Gardner's detractors slammed him as "incapacitated to fill the office from a want of knowledge of the practical duties of his position, and from his advanced age . . . He leaves the control of the office to clerks who are indolent and careless, so that work from the field is frequently unexamined and unapproved for months and sometimes years." Citing other failures, the memorial's authors termed Gardner "tyrannical, arbitrary, and petulant to citizens who apply to him as claimants of the public lands." Of the office staff, the accusers charged, "[Gardner] has brought to Oregon . . . three clerks, Busey, Thompson, and Jones, who are whigs and knownothings [members of the latter group, also known as Nativists, fiercely opposed the large numbers of immigrants flooding into the US from abroad]." Gardner had, the memorialists alleged, "retained in office many of the clerks of his Whig predecessor to the exclusion of able and competent men of the democratic party." Finally, the petitioners charged, "In awarding contracts to deputy surveyors . . . his chief requirement in qualifications has been and now is that the applicant does *not* belong to the Democratic party of Oregon." Entreating the president to act promptly, the Democrats closed, "We, therefore, most respectfully request . . . that Chas. K. Gardner be removed from the office of Surveyor General of Oregon, and that a competent, active national democrat be appointed in his place."[32]

Defending himself, the embattled surveyor general called the charge that he was "tyrannical, arbitrary and petulant" a "bold falsehood" as was the memorial's description of him as "nearly eighty years of age." He denied delaying examination of field notes and defended his selection of clerks and draftsmen. Against the accusation that he retained only Whigs as deputy surveyors, Gardner shot back that the charge was "made out of whole stuff." "The records of the General Land Office show the names of all the principal surveyors, who have been, as they are known to be here, democrats, viz.: Messrs. Butler Ives, George W. Hyde, Wells Lake . . . Harvey Gordon, Lafayette Cartee, Daniel Murphy and his son [Matthew O. C. Murphy] Col. N. Ford, Maj. A. N. Armstrong, Dr. A. W. Patterson . . . besides a son of mine."[33] In response, the *Statesman* editor scoffed, "The list of names given as deputy surveyors contains some of the most arrant political disorganizers. Two of the number were always 'Preston democrats' and when in the field were always advocating *fusion*."[34] Fusion tickets made up of Whig and dissenting, or independent, Democrats appeared in regions outside the farming centers of the Willamette Valley, including Portland, southwestern Oregon, and the coastal counties. Asahel Bush, effectively Oregon's Democratic "boss" during these years, loathed the dissenting Democrats as traitors. The surveyor general's political enemies, the inimitable Bush chief among them, got the president's attention. Charles Gardner had only a few months left in office.[35]

The Donation Land Claim Act expired on December 1, 1855, and Oregon claimants fell subject to other federal laws governing preemption and purchase of public lands.[36] Although opportunities to claim land under the old measure had ended, the land offices at Oregon City, Winchester, and Olympia processed documents for settlers seeking title to land already claimed and handled applications to acquire acreage under the new regulations. In the meantime, fewer qualified deputy surveyors applied for contracts. The US government's reduction of pay rates was, as the surveyor general protested, "ruinous to contractors, if any could be found to undertake the work."[37] "A deputy surveyor," he observed, "however desirous he may be of obtaining work, shrinks from even a hint to take a contract . . . and why? Not that he is afraid of the labor, but because, as he at once tells, 'it won't pay.'" With most of the interior valley surveys complete, the only areas left were far from the central interior valleys, expensive to provision, and choked with dense undergrowth "whose penetration is enough to try the patience and cool the ardor of the most sanguine and persevering surveyor." These conditions, the surveyor general advised the General Land Office, meant that "nearly all the early non-resident deputies of the territory have left the coast. A few lingered among us: their access to and familiarity with information in this office possibly induced them to forgo reasonable salaries in this office to take the last good contracts obtainable in Oregon."[38] Finally, the surveyor general declared, the pool of men available for hire as assistants had been nearly exhausted by ongoing calls for volunteers to "suppress Indian difficulties in Oregon and Washington Territories."

In October, 1855, following the defeat of US Army Major Granville Hallers by Yakima forces in Washington Territory, tensions escalated between Natives and settlers throughout the Oregon Country. The hostilities north of the Columbia River came simultaneously with renewed butchery in the Rogue River Valley, where volunteer militia massacred over two dozen men, women, and children at the mouth of Little Butte Creek. In retaliation, Native avengers murdered miners and settlers along the Rogue River from Rock Point down to Galice. As news of the outbreaks in Washington Territory east of the Cascades and in southwestern Oregon spread, citizens feared an

Indian attack on the Willamette Valley. Like other towns throughout the region, Oregon City formed its own guard for protection. George Hyde was among those volunteering, signing on as a second lieutenant.[39]

As the perceived threat subsided, Hyde assessed his situation. From Illinois, John Preston encouraged him to come home. "My good brother," Preston advised, "make all the money you can . . . and come back to America where there are some people, life and activity. [Illinois] is very prosperous and going ahead and I think money can be used to good advantage here as in Oregon or California — and with more safety."[40] For George Hyde, the Oregon adventure was over. He settled Preston's real estate transactions and packed tracings and notes for the territorial map that the former surveyor general planned to publish. He stopped at O'Neill's store and called on old John McLoughlin. At the last, Hyde said goodbye to his sister, who cried when he went away. He boarded the waiting steamer and, as the pilot maneuvered away from the dock, looked out over the now-familiar Oregon landscape. The vessel navigated the Willamette and Columbia toward Astoria and a steamer that would take him to San Francisco and the long voyage home.

Posterity's Debt to the Surveyors

With George Hyde's departure in 1855, the last of the original government surveyors left Oregon. However, their work, part of a continuum when they arrived, remained so after they left. Although most of the interior western valleys had been surveyed, areas along the coast, in the foothills, the High Cascades and other outlying areas still awaited survey (Figure 56). Government surveyors kept on, extending outward from the grid established by James Freeman, William Ives, and their colleagues. In 1858, Congress expanded the public land surveys to east of the Cascade Mountains in Oregon and Washington territories. That year, surveyors extended the Willamette Base line eastward to a point about twelve miles north of Pendleton and, by late 1860, had carried it nearly to the Snake River. In 1857–1858 the Coast Meridian was surveyed in three separate projects between the

Figure 56. Public Surveys completed 1851–1855. (Map image from Stuart Allan, et al., *Atlas of Oregon.* Eugene, Oregon: University of Oregon Press, 2001.)

vicinity of present-day Waldport and the Oregon-California boundary. Oregon's northern boundary between the Columbia and Snake rivers was surveyed in 1864, and the Oregon-California and Oregon-Idaho boundaries were surveyed in 1867.[41] By the end of the Civil War, the public land surveys extended farther east of the Cascades along the Columbia and Deschutes rivers, in the valley of the John Day River, and in the Umatilla, Grande Ronde, and Powder River valleys.

Between the 1870s and the 1890s, surveyors chained the small valleys, foothills, and adjacent mountains of western Oregon, as well as areas of the Coast Range and central and eastern Oregon. From 1910 through the 1930s, surveys focused on the high country, much of it by then part of the national forests, under US Forest Service management, and most of it not open to bona-fide private homestead entry. Through the rest of the twentieth century, surveyors filled in areas of the Cascades, the southern Coast Range, and areas of south-eastern Oregon.[42] To this day, some of the most rugged, remote high country remains unsurveyed, including areas of the Cascades designated by Congress as wilderness. In the upper Winchuck River drainage near the Oregon coast, whole townships are only now being surveyed by the General Land Office's successor agency, the Bureau of Land Management.[43]

From mid-1851 through late 1855, Oregon's first government land surveyors walked thousands of miles to stake lines in the Chehalis, Cowlitz, Columbia, Willamette, Umpqua, and Rogue River valleys. While the tenure of these surveyors in the Territory was transitory, the wood posts they pounded in precisely measured segments from Puget Sound to the Siskiyous and from the Cascades and the Coast Range were not. These markers signified the peopling of Oregon's western valleys. Counting the Territory's entire population at 12,093 in 1850, the federal census indicates an increase to 52,465 in 1860.[44] As historians Dorothy Johansen and Charles Gates observe, 30,000 to 35,000 persons arrived in Oregon between 1850 and 1855, boosting land sales, increasing the demand for goods and services, and promoting the construction of homes, industries, and whole communities, all of which provided a firm economic foundation for the new state in 1859.[45]

At the same time that they enhanced Euro-American settlement west of the Cascades, the surveyors helped destroy ancient traditions on lands long held sacred by native peoples. The Native inhabitants' resentment of the newcomers was well founded. "Above all else," historian Stephen Dow Beckham has written:

The Indians in the region had been reduced to starvation and to surviving virtually as refugees in their own homeland . . . The settlers' hogs had eaten the acorns and the cattle had cropped off the camas. The pioneers had erected laboriously made split-rail fences and would not let the Indians burn the fields and hillsides as they had done for ages to produce new seed crops or keep down the brush for good hunting . . . Repeatedly the Indians had been driven from their old villages as farmers moved in to file for land under the Donation Land Act.[46]

Although 7,437 people eventually took claims in Oregon encompassing approximately 2,500,000 acres of land, the Donation Land Claim Act did not quell the hunger for free land in Oregon.[47] After the early emigrants had claimed the best agricultural lands in the valleys of the Western Oregon, another wave of settlers pursued ownership of federal lands through homestead laws and other land-disposal authorities. The passage of the Homestead Act in 1862 offered free public lands for settlement, and in Oregon the law applied to both surveyed and unsurveyed federal lands. During the homestead era, settlers sought lands in Eastern Oregon and bought grant lands from railroad and wagon road companies. Large timber companies bought up parcels of timber land from other homesteaders with the intention of consolidating their holdings into larger blocks of land. Because the public land surveys could not keep up with the demand for land, many homesteads were settled by "squatter sovereignty" or preemption rights, with settlers waiting years for surveys to be completed before they could receive clear title to their land. According to historian Stephen Dow Beckham, fifty-eight percent of the homesteads were not patented until the decades from 1900 to 1919.[48]

Beyond their broad contributions to the Territory's development and stability, in what tangible ways, immediate and long-range, did the government land surveyors affect Oregon? In the short term, they provided employment. Twenty deputy surveyors tallied sixty rectangular survey contracts between October, 1851, and December, 1855. Alone or teamed with a colleague, these surveyors hired hundreds of men for their crews. Scores of other men — their names lost to time — assisted the expeditions as packers, cooks, and laborers. Other deputies took sixty-five donation land claim contracts between August, 1852, and October, 1855, hiring their own assistants from nearby farms and towns. Season after season, the surveyors boosted local economies as they bought horses, mules, equipment, food, and liquor in Oregon City, Portland, Salem, Albany, Champoeg, and Jacksonville. Isolated for weeks at a time in distant townships, the deputies patronized local farmers from whom they bought flour, fresh beef, pork, eggs, vegetables, fruit, and butter.

In configuring the land for settlement in the mid-nineteenth century, the first government surveyors also shaped the landscape we see today. Because land claims acquired in Oregon after 1850 had to conform to the rectangular survey system, their boundaries determined the future locations of roads, houses, barns, and, in some cases, towns. As Beckham has observed, "The system left an indelible blueprint. From the beginning of time nature had etched the land in gentle contours. Henceforth straight lines, section corners, and a massive grid system — followed by roads, timber harvest and fields imprinted the landscape."[49]

But beyond the section lines that outlined the farmers' fragments of earth, beyond statehood, beyond the continuum that marks the development of the place we call Oregon, emerges the surveyors' seeing of the landscape and their recording of that vision in field notes and complex, detailed, beautiful maps. They expected government employees and mapmakers to use their notes and sketches; they knew too, that later surveyors would check their calculations in the search for old corners. But they could not have imagined the ways that archaeologists, botanists, geographers, historians, and environmental scientists would one day use their observations to study the past. Nor could they have envisioned how far into the future Chief Clerk's Moore's charge that they give "a full and complete topographical description of the country surveyed, as to every mater of useful information, or likely to gratify public curiosity," would apply. And they could not have imagined that they, through their notes and maps, would be forever associated with their surveys, fulfilling Moore's prediction that "their field work is destined always to remain identified with their respective names and reputations."

The surveyors' field notes and maps became indispensable to the US government and to state and county officials, as well as to commerce and exploration. As geographer William A. Bowen observed, "Without doubt, [the maps] contain the single most comprehensive and accurate cartographic record of the Oregon frontiers . . . [and] include an astonishingly complete view of the physical and cultural landscape."[50] Present-day engineers take the old records in hand to locate markers and to resolve current boundary disputes. Biologists search the maps and accompanying field notes to document historic vegetation and recreate historic landscapes. Anthropologists seek traces of ancient transportation systems and evidence of Native American food sources and village sites; geographers analyze them to understand settlement and agricultural patterns; historians comb notes and maps for the locations of long-vanished farmhouses, barns, roads, ferries, and towns.

While William Ives, James Freeman, and their colleagues provide solid data for the scientists and humanists who research their notes and maps, they also reveal to us places lost to time. They feed our imaginations. Because of the surveyors, those of us who long to look at what once was, who yearn for more than a romantic vision of the old landscape, can turn the telescope around and see into the past.

Epilogue

*"No one in this city appears to know him,
although his face appears to many familiar."*

William and Butler Ives, James Freeman, and George Hyde struck diverging paths into the future, but their lives, intersecting on the Willamette Meridian in the spring of 1851, were forever linked by their years in the west. What became of John Preston and of these surveyors, who, in miles walked, streams crossed, and mountains scaled, organized the Oregon landscape?

William Ives

William Ives' eighteen months in Oregon capped fifteen years as a public land surveyor. He returned to Michigan late in 1852 and, on April 12, 1853, married Sarah Maria Hyde at Grosse Ile, Wayne County, Michigan, where he farmed and sold orchard and garden produce. He served as Supervisor of Monguagon Township from 1860 to 1864 and in 1867, and as Wayne County Surveyor from 1863 to 1865. Sarah Ives died April 26, 1864, following the birth of their fourth daughter. "I am in grief," William Ives wrote his brother, Butler. "I have met with the greatest misfortune of my life." More than two years later, he married Sarah's sister, Lydia Elizabeth Hyde, and the couple had two daughters. William Ives died May 4, 1874, at his Grosse Ile farm.[1]

213

James Freeman

James Freeman's two years in Oregon were followed by over thirty years as a surveyor in California, where he surveyed Mexican and Spanish land grants for the establishment of land ownership rights and took other public land survey contracts. Early in 1861, he patented land in San Luis Obispo County, holding the property as an investment.[2]

On May 24, 1866, Freeman married Sarah P. Young in San Francisco. Ending work as a US government surveyor that year, he maintained a private surveying office on Sacramento Street and later in the "Montgomery Block" in San Francisco, where he worked through the late 1880s. In later years, Freeman, his wife, and daughter lived in Oakland, California. He fell ill in 1893 and entered the Old People's Home at Pine and Pierce streets in San Francisco, where he died at age seventy-nine on June 24, 1895. James Freeman was buried in Laurel Hill Cemetery in San Francisco. Sarah Freeman lived in the East Bay until her death, July 1, 1907.[3]

Butler Ives

Butler Ives was twenty-four when he returned to Michigan. In the spring of 1855, he resumed work as a deputy surveyor there, taking contracts over the next five years. In 1861, Surveyor General John W. North of newly created Nevada Territory named Ives as Deputy Surveyor General. Returning west that spring, Ives extended the Mount Diablo Meridian into Nevada from California and surveyed standard parallels in the Carson Valley area. On May 16, 1863, Ives was appointed commissioner to establish the boundary line between Nevada Territory and the state of California. He and California Surveyor General J. F. Houghton surveyed the northernmost 310 miles of this common boundary.[4]

By 1866, Ives was employed as one of Samuel Montague's principal surveyors for the Central Pacific Railroad and traveled between western Nevada and the Great Salt Lake in Utah. "I am down here," he wrote his brother William, "preparing an outfit for making a reconnaissance for a railroad route from the Sierra Nevada Mountains to Salt Lake, across the Great Basin, for the Central Pacific R.R. Co." Two years later, he wrote his brother from Fort Halleck, near the head of the Humboldt River, "They keep me out in these infernal regions of salt and desolation because I am familiar with the country and don't fear the Indians, which are a bugbear to most people in the country. In fact, I am a sort of vagabond pioneer of the R. R. Co., singled out for difficult jobs, and with a 'cart blanc' to do pretty much as I please."[5] His job as a lead railroad surveyor culminated in the joining of the rails of the Central Pacific and Union Pacific railroads, the first transcontinental railroad, on May 10, 1869, at Promontory Point, northwest of Ogden, Utah.

On Christmas Day, 1871, Ives, forty-one, rode a train from Sacramento to Napa County to assess damage to the line after heavy rains. As the train crossed a bridge four miles from Vallejo, California, he fell from a car, cut his head, and drowned in the tule fields' shallow water. In reporting the incident, the *Sacramento Daily Union* noted that the victim carried his surveying instruments with him when he died. "No one in this city appears to know him," the journalist reported, "although his face appears to many familiar."[6] As a courtesy to the Central Pacific Railroad and in tribute "to Mr. Ives' attainment as R. R. Engineer and character as a gentleman," Wells Fargo and Company forwarded the body to Michigan without charge. Butler Ives' brothers buried him in Elmwood Cemetery in Detroit.[7]

George Hyde

George Hyde's four years in Oregon marked his last as a surveyor. He returned to Joliet, Illinois, early in 1856, and on March 24, 1857, married Mary Howard. The two became parents of five children, one of whom they named George Preston Hyde. The former deputy surveyor invested $15,000 in the flour mill built by his uncle, Henderson Howk, and by John Preston. Taking advantage of Joliet's extended railroad transportation links and the Illinois-Michigan Canal, the trio embarked together on this promising venture. Preston left the firm shortly after the flour mill's completion in 1857. Howk and Hyde continued on as partners until Hyde eventually acquired full interest in the mill.[8]

In 1858, George Hyde still waited for final payment on Contract 54 in Oregon. "A long time has elapsed since the completion of the surveys by Lake & Hyde in Southern Oregon," Lake wrote Surveyor General John Zieber for the two of them, "I am induced to address you a note of enquiry as to what situation the matter is now in . . . [I] am confident you will not allow me to be unfairly dealt by in this matter. I shall esteem it a favor if you will have the matter finally closed up and inform me of the result at an early day."[9]

George Hyde operated the flouring mill until his retirement in 1887. His son recalled that his father "made considerable money in the mill, especially after the war, but lost most of it in outside manufacturing ventures." George Warren Hyde died at the age of sixty-eight on November 29, 1891, and was buried in Oakwood Cemetery.[10]

John B. Preston

John Preston's years in Oregon embellished an already distinguished professional career. He returned to Lockport, Illinois, in the spring of 1855 as superintendent of the Illinois-Michigan Canal and, in 1856, privately published the commercial map based on information compiled on the Oregon surveys. Preston owned a founding interest in the Joliet (Illinois) Flour Mills, served as Secretary of the Chicago and Joliet Railroad, Commissioner of the Illinois State Penitentiary, and, in 1864, opened a wholesale commission business at St. Louis, Missouri. He and Lucy Preston lost the young son born after they left Oregon to scarlet fever in May, 1859.[11] Two younger children, George Hyde Preston and William Trutch Preston, lived to adulthood.

On April 13, 1865, while visiting his parents at Lockport, Illinois, Preston drowned in the Illinois-Michigan Canal. Forty-eight years old, he left his wife and two young sons. The local newspaper praised Preston as "one of the most prominent scientific and business men of Illinois," while another writer observed, "The 15th of April, the first intelligence reached Joliet in the morning . . . of his sudden death, intelligence which would have . . . saddened the community for more than a day, had it not been almost immediately overshadowed by news [of Abraham Lincoln's assassination] which sobered and saddened the nation."[12] In the late 1880s, Lucy Preston left St. Louis for Seattle, Washington, to live near her oldest son, William, and died there on November 8, 1890.

Their eyes on the heavens at dawn, the surveyors found their own highest point on earth. "The best practical method of determining the true meridian . . . is by observing the north star," Davies' surveying text advised. "If this star were precisely at the point in which the axis of the earth, prolonged, pierces the heavens, then, the intersection of the vertical plane passing through it and the place, with the surface of the earth, would be the true meridian."[13]

Notes

Epigraph

[1] Wallace Stegner, "History Comes to the Plains," The Wallace Stegner Collection, Mariott Library, University of Utah, Salt Lake City, Utah.

Introduction

[1] Glossaries of BLM Surveying and Mapping Terms, Second Edition, 1980 (http://www.fairview-industries.com/gismodule/PartOneHistory.html).

[2] Paul W. Gates, *History of Public Land Law Development* (Washington, DC: Government Printing Office, 1968), 65.

[3] Gates, *The Farmer's Age: Agriculture 1815–1860,* vol. 3, *The Economic History of the United States* (New York, New York: Holt, Rinehart and Winston, 1960), 54.

[4] Thomas D. Clark, *Frontier America: The Story of the Westward Movement* (New York, New York: Charles Scribner's Sons, 1969), 475,652.

[5] Dorothy O. Johansen, "The Roll of Land Laws in the Settlement of Oregon," *Genealogical Material in Oregon Donation Land Claims.* vol. I. (Portland, Oregon: Genealogical Forum of Portland, 1957).

[6] Wallace Stegner, *Beyond the Hundredth Meridian: John Wesley Powell and the Second Opening of the West* (Boston, Massachusetts: Houghton Mifflin, 1954), 213.

[7] *Instructions to the Surveyors General of Public Lands of the United States, For Those Surveying Districts Established in and Since the Year 1850; Containing Also A Manual of Instructions to Regulate the Field Operations of Deputy Surveyors* (Washington, DC: A. O. P. Nicholson, Public Printer, 1855), 35.

Chapter 1

[1] Butler Ives, Diaries, Folder 1, Oct. 1850–June 1851; May–Aug. 1851, Butler Ives Diaries, 1851–1853;1855, 1860. MSS 3011, Container 1, Folders 1, 2, and 3,Western Reserve Historical Society, Cleveland, Ohio. Further reproduction is permissible only with written permission from the Western Reserve Historical Society.

[2] Ibid.

[3] Ibid. May 5, 1851.

[4] William Ives, May 3, 1851, William Ives Correspondence and Papers, 1847–1864, MS 1126, Burton Historical Collection, Detroit Public Library; Butler Ives, May 3, 1851.

[5] William A. Bowen, *The Willamette Valley: Migration and Settlement on the Oregon Frontier* (Seattle, Washington: University of Washington Press, 1978), 13, 15.

[6] Stephen Dow Beckham, *Land of the Umpqua: A History of Douglas County Oregon* (Roseburg, Oregon: Douglas County Commissioners 1986), 161. The full title of the Donation Land Act of September 27, 1850 (9 Statute 496) is "An Act to Create the Office of Surveyor-

General of the Public Lands in Oregon, and to Provide for the Survey and to Make Donations to Settlers of the Said Public Lands."

7 *Senate Executive Journal*, Saturday, September 28, 1850; Monday, December 16, 1850, American Memory, A Century of Lawmaking for a New Nation: US Congressional Documents and Debates, 1774–1875, http://memory.loc.gov/ghi-bin/.

8 *History of Will County, Illinois* (Chicago, Illinois: Wm. Le Baron, Jr. & Co., 1878), 307,421; http://www.rootsweb.com/~ilwill/1878bios/ query025.htm (19 April 2004). John Preston was the oldest son of Isaac and Lovica Walker Preston.

9 The Reverend Calvin Colton quoted in Michael Kimmel, *Manhood in America: A Cultural History* (New York, New York: Simon and Schuster, The Free Press, 1996), 26.

10 *Joliet Republic and Sun* (Joliet, Illinois) vol. 46, no. 17:3, December 4, 1891; N. de Bertrand Lugrin, *The Pioneer Women of Vancouver Island 1843–1866* (Victoria, British Columbia: Women's Canadian Club of Victoria, 1928), 306; "Biographical Sketch of George Warren Hyde," Reed Hyde Family Papers, Private collection. Lucy Hyde was born December 26, 1816, to Louis and Fannie Howk Hyde. Julia Hyde was born September 7, 1825.

11 Illinois and Michigan Canal Document 12: "List of Appointed Canal Officers and Agents," http://www.sos.state.il.us/departments/archives/i&mpack/i&mdoc12.html; "Biographical Sketch of George Warren Hyde," Reed Hyde Family Papers.

12 R. A. Payne to John B. Preston, January 23, 1851, vol. 2, p. 62, Series 3, Miscellaneous Letters Received 1851–1853, Record Group 49, Records of the Bureau of Land Management (RG 49), National Archives and Records Administration — Pacific-Alaska Region, Seattle, Washington (NARA-Seattle).

13 John B. Preston to R. A. Payne, January 29, 1851, vol. 2, p. 81, Series 3, RG 49, NARA-Seattle.

14 William A. Burt, *A Key to the Solar Compass, and Surveyor's Companion* (Philadelphia, Pennsylvania: William S. Young, 1855), 9.

15 John S. Burt, "The Search for an Ancestor: Pioneer Surveyor William A. Burt," *The California Surveyor* (Summer, 1981):11–12; Roy Minnick, *A Collection of Original Instructions to Surveyors of the Public Lands 1815–1881* (Rancho Cordova, California: Landmark Enterprises, c. 1992), 233; C. Albert White, "The Public Land Surveys in Oregon," n.d., Professional Land Surveyors of Oregon, http://www.plso.org/readingroom/GLO.htm.

16 Preston to Butterfield, September 1, 1851; September 30, 1851; Series 2, Letters Sent to Commissioner, General Land Office 1851–1921, vol. A, 1851–1859, 9, 16; Record Group 49, Records of the Bureau of Land Management, Oregon (RG 49); National Archives and Records Administration–Pacific Alaska Region (Seattle); NARA-Seattle; White, "The Public Land Surveys in Oregon."

17 Evans to Preston, March, 1851, vol.1, Item 12a; Series 3, RG 49, NARA-Seattle. Evans arrived in Oregon in late 1851 and returned to Washington, DC, in the fall of 1852. Most of his geological notes made during this tour were lost.

18 Preston to Butterfield, July 1, 1851, Series 2, p. 5, RG 49, NARA-Seattle; White, "Willamette Initial Point History," *The Oregon Surveyor* 21, no. 3 (1993):17; White, *A History of the Rectangular Survey System* (Washington, DC: US Department of the Interior, Bureau of Land Management, 1983) 115; White, *A Casebook of Oregon Donation Land Claims*, 19 (LLM Publications, 2001); White, "The Public Land Surveys in Oregon."

19 White, "Willamette Initial Point History," *The Oregon Surveyor* 21, no.3 (1993): 17; White, "The Public Land Surveys in Oregon." Revised editions of the manual were published in 1855, 1881, 1890, 1894, 1902, 1930, 1947, and 1973.

20 Wallace Stegner, *Beyond the Hundredth Meridian: John Wesley Powell and the Second Opening of the American West* (Boston, Massachusetts: Houghton Mifflin Company, 1953) 213.

[21] Charles Davies, *Elements of Surveying and Navigation: With a Description of the Instruments and the Necessary Tables* (New York, New York: A. S. Barnes & Co., 1847), 126.

[22] Marie E. Gilchrist (ed.) "William Ives' Huron Mountains Survey, 1846," *Michigan History* 50, no. 4 (1966): 323; Gilchrist, "A Michigan Surveyor — William Ives," *Inland Seas* 21 (Winter 1965): 321; William Ives was the fifth of ten children of Butler and Olive Morse Ives.

[23] Gilchrist, "Isle Royale Survey," Part I, *Inland Seas* 24, no. 3 (1968):185; Norman C. Caldwell, *Surveyors of the Public Lands in Michigan 1808–2000* (Owosso, Michigan: Norman C. Caldwell, 2001), 218; Gilchrist (ed.), "William Ives Huron Mountains Survey 1846," 323; White, *A History of the Rectangular Survey System*, 102.

[24] Gilchrist (ed.), "William Ives' Huron Mountains Survey, 1846," *Michigan History* 50, no. 4 (1966): 340. Butler and Olive Ives both died of sudden illness on August 31, 1846.

[25] Gilchrist, "Isle Royale Survey,"184; William Ives Correspondence and Papers, MSS 1126.

[26] Butler Ives, November 27; December 5, 1850, MSS 3011; Butler Ives Surveying Contracts & Appointments, 1851–1861, Stanford University Libraries, Stanford Auxiliary Library 3; "Drives 'F,' 'G,' and 'H'," http://www.umich.edu/~bhl/bhl/huron/drivefgh.html (February 11, 2004); *Evangelist,* April 24, 1879; *Saline Observer*, January 8, 1948, Vertical Files, Bentley Historical Library, University of Michigan, Ann Arbor. Classical languages scholar Dr. Rufus Nutting opened the Lodi Plains Academy in 1847 after his resignation as principal of the Romeo Branch of the University of Michigan.

[27] Butler Ives, March 10–12, 1851, MSS 3011.

[28] Pleasantville Cemetery Records, Fenton History Center Museum & Library, Jamestown, New York. James Freeman's father, Jonathan Freeman, was a native of Connecticut and veteran of the War of 1812. He married Nancy Youngs in 1813 and moved to western New York. Nancy Freeman was born in 1788 and died December 23, 1845. The Freemans are buried in Pleasantville Cemetery on the Hartfield Stockton Road near Chautauqua, New York.

[29] Norman Caldwell, notes to author, April 30, 2001; "History of Potosi," Potosi Township Historical Society, http://www.vangrafx.com/PTHS/potosi/orgin.html; Arunah Parker Papers 1843–1853, Wisconsin Historical Society Archives Catalogue, http://arcat.library.wisc.edu/; Rod Squires, "The Public Land Survey in Minnesota Territory, 1847–1852," http://www.geog.umn.edu/faculty/squires/.

[30] Squires, "The Public Land Survey in Minnesota Territory, 1847–1852."

[31] US Federal Census, 1850, Grant County, Wisconsin, 24th District, 025 (Microfilm Roll Number 444985); James Freeman owned 360 acres in Grant County.

[32] Thomas Nelson to Cornelia Nelson, March 20, 1851, Thomas Nelson Letters from Oregon 1851–1853, WA MSS S-2176, Yale Collection of Western Americana, The Beinecke Rare Book and Manuscript Library, Yale University; Stephen Fowler Steven Fowler Chadwick, "Oregon Party," WA MSS 71, Yale Collection of Western Americana, The Beinecke Rare Book and Manuscript Library, Yale University; Louis J. Rasmussen, *San Francisco Ship Passenger Lists*, vol. 2 (Baltimore, Maryland: Deford and Company, 1966), 131–132.

[33] William W. Glenn and Jeanne E. Glenn, "Early Public Land Surveyors in the Northwest," *The Oregon Surveyor*, 18 (July–August 1990): 21. Moody, a son of Major Thomas H. and Hanna Ferry Moody, was born May 27, 1832, in Massachusetts. Allan Seymour was born November 17, 1833. His half-sister, Cornelia Seymour, married Thomas Nelson on June 4, 1844. Seymour's first name is spelled variously as Allen or Allan.

[34] Cedric Ridgley-Nevitt, *American Steamships on the Atlantic,*" (Newark, Delaware: University of Delaware Press, 1981), 121–125; Erik Heyl, *Early American Steamers*, vol. 1 (Buffalo, New York: Erik Heyl, 1953), 141.

35 Verda Spickelmier, "The Oregon Territory's First Surveyor General Left His Mark," Oregon City *Enterprise Courier,* December 15, 1991, section 3A; Oregon City *Oregon Spectator*, March 13, 1851, 2:6; E. Ruth Rockwood, ed., "Diary of G. H. Atkinson, 1847–1858," *Oregon Historical Quarterly* 41 (1940): 212; James R. Perry, Richard H. Chused, and Mary De Lano (eds.) "The Spousal Letters of Samuel Royal Thurston, Oregon's First Territorial Delegate to Congress 1849–1851,"*Oregon Historical Quarterly* 96 (1995): 62; The teachers were Mary Almira Gray of Vermont, Elizabeth Lincoln of Maine, and Margaret Wands, Sarah Smith, and Elizabeth Millar of New York.

36 Thomas Nelson to Cornelia Nelson, March 29, 1851, WA MSS S-2176.

37 Preston to Butterfield, March 31, 1851, John B. Preston Collection, MSS 914, Oregon Historical Society Research Library, Portland, Oregon.

38 Thomas Nelson to Cornelia Nelson, April 9, 1851, April 23, 1851, WA MSS S-2176.

39 Ibid., April 27. 1851.

40 Schwantes, Carlos Arnaldo. *Long Day's Journey: The Steamboat & Stagecoach Era in the Northern West (*Seattle, Washington: University of Washington Press, 1999), 81; Oregon City *Oregon Spectator,* May 1, 1851, 3:2; 3:3; May 8, 1851, 3:3; Oregon City *Oregon Statesman*, May 2, 1851, 2:4.

41 *Oregon Statesman*, May 6, 1851.

42 Howard McKinley Corning, *Willamette Landings: Ghost Towns of the River (*Portland, Oregon: Binfords and Mort, 1947), 32–33; *Oregon Statesman,* May 2, 1851, 2:4.

43 Corning, *Willamette Landings, Ghost Towns of the River*, 32–33.

44 Thomas Nelson to Cornelia Nelson, May 6, 1851, WA MSS 2176.

45 Rockwood, ed., "Diary of G. H. Atkinson 1847–1858," 212; Stephen Fowler Chadwick, Oregon Party, WA MSS 71. The teachers assumed their jobs and soon married: Margaret Wands to Governor John P. Gaines, Sarah Smith to Alanson Beers, Almira Gray to B. F. McLench, Elizabeth Lincoln to A. A. Skinner, and Elizabeth Millar to Joseph Gardner Wilson.

46 John W. Reps, *Cities of the American West: A History of Frontier Urban Planning* (Princeton, New Jersey: Princeton University Press, 1979), 346; George McFall, Journal of George McFall, September 21, 1851. Typescript held in the Jackson County, Oregon, Surveyor's Office.

47 Thomas Nelson to Cornelia Nelson, May 21, 1851, WA MSS 2176.

48 Dorothy O. Johansen and Charles M. Gates, *Empire of the Columbia: A History of the Pacific Northwest* (New York, New York: Harper and Row, 1967), 212; Reps, *Cities of the American West*, 346, 348; *Oregon Spectator*, February 13, 1851, 2:1; William Allen Bentson, "Historic Capitols of Oregon: An Illustrated Chronology," Salem, Oregon, Oregon Library Foundation, http://www.osl.state.or.us/lib/capitols/index.htm.

49 Thomas Nelson to Cornelia Nelson, May 21, 1851, WA MSS 2176.

50 Johansen and Gates, *Empire of the Columbia*, 243.

51 Preston to Butterfield, May 5, 1851, MSS 914; Preston to Butterfield, September 1, 1851, Series 2, p. 9, RG 49, NARA-Seattle; *Oregon Statesman,* May 9, 1851, 2:4.

52 White, *History of the Rectangular Survey System*, 18, 83; Burt Brown Barker, "The Estate of Dr. John McLoughlin: The Papers Discovered," *Oregon Historical Quarterly*, 50, no. 3 (1949): 157. The building that housed the Surveyor General's Office stood on Block 29, Lot 8.

53 Hugh Goldsborough to John Preston, April 12, 1851; Series 3, vol 1, p. 88, RG 49, NARA-Seattle.

54 White, "The Public Land Surveys in Oregon."

[55] White, "Willamette Initial Point History" *The Oregon Surveyor* 21, no. 3 (1993):19–20. Preston to Butterfield, May 5, 1851, MSS 914; The 1851 *Manual of Instructions* describes a random line as a "trial line" on which the surveyor would set "*temporary* mile and half-mile stakes."

[56] William Ives, March 17, 1851, MSS 1126; Norman C. Caldwell, *Surveyors of the Public Lands in Michigan 1808–2000,* 39–40, 208–210. Joseph Hunt worked four seasons for Ives. Brevoort, brother of Michigan surveyor Henry Brevoort, had years of experience on survey crews.

[57] Charles Noble to John Preston, March 15, 1851, Series 5, Box 120, vol. 1, Item No. 2; "Letters Received from Deputy Surveyors 1851–1879," Record Group 49 (RG 49) NARA Pacific-Alaska Region, Seattle; Bela Hubbard to John Preston, March 15, 1851, Series 5, Box 120, RG 49, "Letters Received from Deputy Surveyors," Item 4; William Burt to John Preston, March 11, 1851, Series 5, Box 120, "Letters Received from Deputy Surveyors," vol. 1, Item No. 1, NARA-Seattle.

[58] Heyl, vol. I *Early American Steamers*, vol. 1, p. 77; William Ives, March 25, 1851, MSS 1126; Butler Ives, March 29, 1851, MSS 3011.

[59] Butler Ives, April 6, 1851, MSS 3011.

[60] Ibid., April 9–10, 1851.

[61] William Ives, April 15, 1851, MSS 1126; Butler Ives, April 20, 1851, MSS 3011.

[62] Butler Ives, April 23, 1851, MSS 3011.

[63] Ibid., May 6, 1851.

[64] Ibid., May 10, 1851.

[65] Ibid., May 11, 1851.

[66] Ibid.

Chapter 2

[1] Butler Ives, May 20, 1851, MSS. 3011.

[2] Ibid., May 14–16, 1851; http://web4.si.edu/surveying/ (August 26, 2002). Burt's solar apparatus has three arcs. One is to set the latitude of land to be surveyed, another is to set the declination of the sun, and the third is to set the hour of the day.

[3] Butler Ives, May 20, 1851, MSS 3011; William Ives, May 20, 1851, MSS 1126.

[4] Butler Ives, May 23, 1851, MSS 3011.

[5] Ibid., May 23–24, 1851.

[6] Ibid., May 28, 1851; William Ives, May 28, 1851, MSS 1126; Contracts 1 and 2, Series 22, Contracts and Bonds 1851–1870, Record Group 49, Records of the Bureau of Land Management (Records of the Surveyor General of Oregon and the Oregon Cadastral Survey Office); (NARA-Seattle).

[7] Lowell O. Stewart, *Public Land Surveys* (New York, New York: Arno Press, 1979), 51.

[8] Contracts 1 and 2, Series 22; RG-49; NARA-Seattle.

[9] Ibid.; White, "The Public Land Surveys in Oregon."

[10] Bud [Francois D.] Uzes, *Early Survey Instructions, The Compleat Surveyor — Books & Instruments*, http://www.uzes.net/books/bookreviews/surveyinginstructions.htm (April 20, 2004); White, *A History of the Rectangular Survey System*, 291, 359.

[11] Roy Minnick, *A Collection of Original Instructions to Surveyors of the Public Lands, 1851–1881* (Rancho Cordova, California: Landmark Enterprises, 199, 225.

[12] Ibid., p. 254.

[13] Norman Caldwell, e-mail to author, July 22, 2002.

[14] Minnick, 254–256.

[15] Francois D. Uzes, *Chaining the Land: A History of Surveying in California* (Sacramento, California: Landmark Enterprises, 1977), 18, 164.

[16] White, *A History of the Rectangular Survey System*, 114.

[17] Contract 1, Series 22, RG 49, NARA-Seattle.

[18] Davies, *Elements of Surveying and Navigation*, 127.

[19] Butler Ives, May 18, 1851, MSS 3011.

[20] *Oregon Journal,* June 4, 1916; Butler Ives, June 1, 1851, MSS 3011.

[21] BLM Field Notes, vol. OR-R0002, p. 0129.0, Township 1 North, Range 1 East, 1851 (West Boundary), http://www.or.blm.gov/or/landrecords/survey/ySrvy.php (October 12, 2007). Nicholas Coarser's name is spelled variously as Courser, Corsair, and Coarser.

[22] William Ives, April 10, 1852, Diary and Expense Account, March 14, 1852–November 29, 1852, MSS 3011; Preston to Butterfield, June 14, 1851, John B. Preston Collection, MSS 914.

[23] BLM Field Notes, vol. OR-0002, p. 0129.0. Township 1 North, Range 1 East, 1851 (West Boundary).

[24] Butler Ives, June 4, 1851, MSS 3011; William Ives, June 4, 1851, MSS 1126.

[25] The initial point for the Mount Diablo Meridian in California was set in July, 1851. The earliest initial point (Washington Meridian) was set in 1803 in Mississippi, the latest (Kateel River and Umiat Meridians) in Alaska in 1956. Some states have more than one principal meridian; others have only one. Eight public land surveys performed between 1785 and 1805 have no initial point as an origin for township identification (*Manual of Instructions for the Survey of the Public Lands of the United States*, Bureau of Land Management Technical Bulletin 6, US Department of the Interior, 1973).

[26] By 1885, the cedar post had decayed and the bearing trees had been logged. That year, surveyors placed a limestone marker at the site, and in later years this marker was known as the "Willamette Stone." In the 1940s, approximately 1.6 acres were set aside to create Willamette Stone State Park. Recurring vandalism between 1950 and 1987 led to the installation of a stainless steel monument to replace the limestone marker. For additional information on the history of the Willamette Stone see White, "Willamette Initial Point History," *The Oregon Surveyor* 22, no. 1: 18–20.

[27] William Ives, June 4, 1851, MSS 1126; Butler Ives, June 4, 1851 MSS 3011; Contracts 1 and 2, Series 22, RG 49, NARA-Seattle; Norman Caldwell, communication to author, June 26, 2004. Caldwell observes that surveyors and mathematicians recognize the theory of "multiple measurement" today.

[28] Norman Caldwell, July 22, 2002, e-mail to author; Minnick, 255.

[29] These points included township and section corners. Using a scribe, the axeman marked the appropriate number of the township or section according to the direction faced by each of the posts' elevations.

[30] Minnick, 235.

[31] Contracts 1 and 2, Series 22, RG 49, NARA-Seattle.

[32] Minnick, 234–235.

[33] White, "The Public Land Surveys in Oregon."

[34] Ibid.; Minnick, 234.

[35] Minnick, 234.

[36] Contracts 1 and 2, Series 22; RG 49, NARA-Seattle; BLM Field Notes, vol. OR-R0002, pp. 0128.0–0129.0, Township 1 North, Range 1 East, 1851 (West Boundary).

[37] Butler Ives, June 4, 1851, MSS 3011; William Ives, BLM Field Notes, Willamette Meridian Survey, vol. OR-R0002 p. 0138.0 Township 1 North, Range 1 East (West Boundary). Milton Doane settled in the Willamette Valley in the mid-1840s; White, "Willamette Initial Point History," *The Oregon Surveyor* (July/August 1993): 11.

[38] Butler Ives, June 6, 1851, MSS 3011; Contracts 1 and 2, Contracts and Bonds 1851–1870, Series 22, Box 48, RG 49, NARA-Seattle.

[39] *Portland Oregonian,* June 7, 1851, quoted in White, "Willamette Initial Point History," *The Oregon Surveyor* 21, no. 4 (July/August 1993): 8.

[40] Butler Ives, June 7–8, 1851, MSS 3011.

[41] William Ives, BLM Field Notes, Willamette Meridian, vol. OR-R0002, p. 0172.0, Township 3 North, 1 West, W.M. (East Boundary).

[42] Butler Ives, June 9, 1851, MSS 3011.

[43] Ibid., June 10–11, 1851; William Ives to John B. Preston, June 13, 1851, Series 5, vol. 1, RG 49, NARA-Seattle.

[44] Butler Ives, June 11, 1851, MSS 3011; William Ives to Preston, June 13, 1851, Series 5, vol. 1; RG 49, NARA-Seattle. Walter Pomeroy settled a donation land claim in Township 1 North, Range 2 West encompassing parts of Sections 21, 22, 27, and 28.

[45] William Ives to John Preston, June 13, 1851, Series 5, vol. 1, RG 49, NARA-Seattle; William Ives, BLM Field Notes, vol. OR-R0002, p. 0076.0, Township 1 North, Range 1 East (South Boundary); F. H. McNeil, "Official Survey of Oregon Country Was Undertaken 65 Years Ago Today,"*Oregon Sunday Journal* (Portland, Oregon), June 4, 1916, 2: 9.

[46] William Ives to John Preston, June 13, 1851, Series 5, vol. 1, RG 49, NARA-Seattle.

[47] Butler Ives, June 17, 1851, MSS 3011.

[48] William Ives, June 17, 1851, MSS 1126.

[49] Minnick, 226.

[50] Butler Ives, June 18, 1851, MSS 3011. The base line ran just south of the academy building then under construction.

[51] Ibid., June 18–19, 1851, MSS 3011. Allen's mill stood near Gale's Creek in Township 1 North, Range 4 West, Section 34.

[52] Butler Ives, June 21–22, 1851, MSS 3011.

[53] Ibid., June 23–26, 1851.

[54] William Ives to John Preston, September 6, 1851, Series 5, vol. 1, RG 49, NARA-Seattle; BLM Field Notes, vol. OR-R0002 p. 0065.0, Township 1 North, 6 West (West on South Boundary).

[55] BLM Field Notes, vol. OR-R0002, p. 0062.0 Township 1 North, Range 6 West (South boundary). Ives stood on the base line along the south boundary of Section 32 in Range 6 West.

[56] Katherine C. Atwood and Frank A. Lang, *As Long as the World Goes On: Environmental History of the Evans Creek Watershed* (USDI Bureau of Land Management, Medford District Office, 1995), 149–150; Jeff LaLande, personal communication, August 17, 2004.

[57] BLM Field Notes, vol. OR-R0002, pp. 0065.0–0066.0.

[58] Butler Ives, July 27–30, MSS 3011.

[59] William Ives to Preston, July 1, 1851, Series 5, vol. 1, RG 49, NARA-Seattle.

[60] Ibid.

[61] BLM Field Notes, vol. OR-R0001 p. 0111.0 Township 14 South, Range 1 East (West Boundary); vol. OR-R0001, p. 0272.0, Township 30 South, Range 5 West, North

Boundary; C. Albert White, "Harvey Gordon: US Deputy Surveyor and Designer of the Oregon State Seal," *The Oregon Surveyor* 35, no. 2 (2002): 22.

[62] Contract 1, Series 22; RG 49 SG/CS; NARA-Seattle; *Oregon Spectator*, November 18, 1851, 2:3.

[63] James Freeman, "Synopsis of the Survey of Willamette Meridian to Umpqua," n.d. (c. September 1, 1851), Series 5, vol. 1, RG 49, NARA-Seattle. The first standard parallel was the southern boundary of Township 5 South; Minnick, p. 230. The 1851 *Manual of Instructions* stipulated that north of the Columbia River, standard parallels would "run at distances of every four townships, or twenty-four miles" and that south of the river, standard parallels would be established "at distances of every five townships, or thirty miles." Surveyors established standard parallels at planned intervals to counteract error resulting from the converging of meridians.

[64] Freeman to Preston, June 20, 1851, Series 5, vol. 1, RG 49, NARA-Seattle.

[65] Thomas Nelson to Cornelia Nelson, July 21, 1851, WA MSS S-2176.

[66] Minnick, pp. 256–257; White, "Willamette Initial Point History," *The Oregon Surveyor* 22, no. 1 (1994): 19. White observes that these geodetic notes are copies of the regular notes with some ties to topographic features and claimants' homes. They have, he says, "apparently never served any significantly useful purpose;" Kimball Webster, *The Gold Seekers of '49: A Personal Narrative of the Overland Trail and Adventures in California and Oregon from 1849 to 1854*. (Manchester, New Hampshire: Standard Book Company, 1917), 203.

[67] Freeman to Preston, c. September 1, 1851, Series 5, vol. 1, RG 49, NARA-Seattle.

[68] BLM Field Notes, vol. OR-R0001, pp. 0316.0–0319.0, Township 11 South, Range 1 East (West Boundary).

Chapter 3

[1] White, *A Casebook of Oregon Donation Land Claims*, 21–22; *Oregon Spectator*, June 21, 1851; June 27, 1851, 3:1.

[2] Preston to Butterfield, July 1, 1851 Series 2, vol. A, 1851–1859, p. 5, RG 49, NARA-Seattle.

[3] *Oregon Spectator,* July 3, 1851, 3:1; June 19, 1851 2:1, 2:2; July 26, 1851 2:3; July 29, 1851 2:1; *Portland Oregonian,* July 5, 1851 2:3; Oregon City *Oregon Statesman*, July 22, 1851 2:7.

[4] Jesse Applegate to John Preston, May 8, 1851, Series 3, Box 49, vol. 2, item 48, RG 49, NARA-Seattle; Preston to Butterfield, July 23, 1851, Series 3, Box 48, item 7, RG 49, NARA-Seattle; Jesse Applegate to John Preston, n.d. Series 3, Box 49, vol. 2, item 40, RG 49, NARA-Seattle.

[5] William Ives to John Preston, July 1, 1851; Series 5, vol. 1, RG 49, NARA-Seattle.

[6] Judy Card, "Early Families in the Woodland Community: A 1958 Report of the Woodland History Committee," http://www.lewisriver.com/pt2-page2.html; Samuel Gatton's brother, William, settled near the present site of St. Johns in Multnomah County, Oregon. Gatton Creek, which flows into Ramsey Lake north of St. Johns, is named for the pioneer settler.

[7] Grant Nelson, "The Early Years of Mt. Tabor," Mt. Tabor Neighborhood Association Newsletter (n.d.), http://www.mttaborpdx.org/history_early_years.html; Butler Ives, July 2–3, 1851, MSS 3011; BLM Field Notes, vol. OR-R0002, p. 0079.0–p. 0080.0, Township 1 North, Range 1 East (South Boundary).

[8] Butler Ives, July 7, 1851, MSS 3011.

[9] Ibid., July 7–8, 1851.

[10] BLM Field Notes, vol. OR-R0002, pp. 0125.0–0127.0, Township 1 North, Range 6 East (South Boundary); White, "Willamette Initial Point History," *The Oregon Surveyor* 21, no. 4 (1993): 17. White notes that the actual summit "would be about 7 or 8 miles farther east of Ives' stopping point."

[11] Butler Ives, July 12, 1851, MSS 3011.

[12] James Freeman, "Synopsis of the Survey of Willamette Meridian to Umpqua," c. September 1, 1851, Series 5, Box 120, vol. 1, RG 49, NARA-Seattle.

[13] Ibid. The surveyor traveled west along the line between Townships 13 and 14 South to the township corner between Ranges 3 and 4 West. Townships continued to be counted from the true meridian line.

[14] BLM Field Notes, vol. OR-R0001 pp. 0328.0–0329.0, Township 14 South, Range 1 West (East Boundary); Freeman to Preston, c. September 1, 1851, "Synopsis of the Survey of Willamette Meridian to Umpqua," Series 5, Box 120, Vol. 1, RG 49, NARA-Seattle.

[15] Peter Boag, *Environment and Experience: Settlement Culture in Nineteenth Century Oregon* (Berkeley, California: University of California Press, 1992), 46; 123.

[16] Contract No. 1, Series 22, RG 49, NARA-Seattle; BLM Field Notes, vol. OR-R0001 p. 0329.0, Township 14 South, Range 1 West (East Boundary).

[17] Freeman to Preston, July 20, 1851, Series 5, Box 120, vol. 1; RG 49, NARA-Seattle. Freeman inverted Stewart's name in his report.

[18] For a thorough assessment of the social and economic impact of settlement in the Calapooia Valley, see Peter G. Boag's *Environment and Experience: Settlement Culture in Nineteenth-Century Oregon*, (Berkeley, California: University of California Press, 1992).

[19] Preston to Butterfield, July 23, 1851, Series 2, vol. A, p. 7, RG 49, NARA-Seattle. Preston asked the Commissioner's permission to consider the offsets as standard parallels; Preston to Butterfield, June 14, 1851, Series 2, vol. A, p. 4, RG 49, NARA-Seattle. Preston expected to start the township surveys before the deputies had finished the meridian and base line surveys.

[20] Thomas Nelson to Cornelia Nelson, July 21, 1851 WA MSS 2176. Referring to the surveyor general's recent time on the road, Nelson noted, "Mr. Preston . . . saw Allan a short time since . . . Mr. Freeman is much pleased with Allan and expressed himself well satisfied with the manner in which he discharges his duties." Preston to Butterfield, July 23, 1851, Series 2, Box 48, Book A, p. 7, RG 49, NARA-Seattle.

[21] A native of West Virginia, William Vaughn was related to Nancy Vaughn Ferguson.

[22] Freeman to Preston, July 20, 1851, Series 5, vol. 1, RG 49, NARA-Seattle.

[23] BLM Field Notes, vol. OR-R0001, p. 0186.0, Township 19 South, Range 3 West (West Boundary).

[24] James Freeman, "Synopsis of the Survey of Willamette Meridian to Umpqua," c. September 1, 1851, Series 5, vol. 1, RG 49, NARA-Seattle. The USGS Quadrangle Map, Harness Mountain 7.5 minute series (1987) gives an elevation of 2,462 feet for Hobart Butte. The surveyor set the corner for the corner of Townships 22 and 23 South, Ranges 3 and 4 West.

[25] BLM Field Notes, vol. OR-R0001, pp. 0207.0–0208.0, Township 23 South, Range 4 West (West Boundary).

[26] James Freeman, "Synopsis of the Survey of Willamette Meridian to Umpqua," c. September 1, 1851; Series 5, vol. 1, RG 49, NARA-Seattle; Stephen Dow Beckham, *Land of the Umpqua, A History of Douglas County, Oregon*, 75.

[27] BLM Field Notes, vol. OR-R0001, p. 0231.0, Township 24 South, Range 4 West (West Boundary), vol. OR-R0001, p. 0.232.0, Township 25 South, Range 4 West (West Boundary).

28 James Freeman, "Synopsis of the Survey of Willamette Meridian to Umpqua," c. September 1, 1851; Series 5, vol. 1, RG 49, NARA-Seattle; BLM Field notes OR-R0001 p. 0240.0 Township 26 South, Range 4 West (West Boundary); OR R-0001 p. 0251.0 Township 27 South, Range 4 West (Western Boundary); vol. OR-R0001 p. 0255.0, Township 28 South, Range 4 West (West Boundary).

29 BLM Field Notes, vol. OR-R0001, p. 0270.0–0271.0, Township 30 South, Range 4 West (West Boundary), August 12, 1851.

30 James Freeman, "Synopsis of the Survey of Willamette Meridian to Umpqua," n.d. (c. September 1, 1851); Series 5, vol. 1, RG 49, NARA-Seattle.

31 Applegate to Preston, August 21, 1851, Series 3, vol. 2, item 37, RG 49, NARA-Seattle.

32 Butler Ives, July 12; 17, 1851, MSS 3011.

33 William Ives to John Preston, July 22 1851, Series 5, vol. 1, RG 49, NARA-Seattle.

34 BLM Field Notes, vol. OR-R0002, p. 0205.0, Township 4 North, Range 1 East (West Boundary). Joseph Lambert arrived in Oregon in 1850 after a brief stint at mining in California. He later settled a donation land claim and became well known as a horticulturist and developer of the Lambert cherry.

35 Butler Ives, July 18, 1851, MSS 3011; *Biographical Directory of the United States Congress 1774–Present*, http://bioguide.congress.gov/scripts/. Lancaster ran unsuccessfully as a delegate to the 31st US Congress from Oregon but was later elected, serving from April 12, 1854, to March 3, 1855. Butler Ives devises his own spelling of Cathlapootle, an early name for the river. Edward S. Meany, author of *Origin of Washington Geographic Names,* (Seattle, Washington: University of Washington Press, 1923), 146, notes that the river was later named for area land claimant Adolphus Lee Lewis.

36 Butler Ives, July 19, 1851, MSS 3011.

37 BLM Field Notes, vol. OR-R0002, p. 0231.0, Township 5 North, Range 1 West (East Boundary).

38 William Ives to John Preston, September 6, 1851; Series 5 vol. 1, RG 49, NARA-Seattle; Butler Ives, July 20–21, 1851, MSS 3011.

39 William Ives to John Preston, July 22, 1851; Series 5, vol. 1, RG 49, NARA-Seattle.

40 Ibid., July 22, 1851; September 6, 1851.

41 Butler Ives, July 22, 1851, MSS 3011.

42 Ibid., July 23–24, 1851.

43 BLM Field Notes, vol. OR-R0002, p. 0240.0, Township 6 North, Range 1 West (East Boundary).

44 BLM Field Notes, vol. OR-R0002, p. 0244.0, Township 7 North, Range 1 West (East Boundary).

45 Butler Ives, July 25, 1851.

46 Ibid., July 26, 1851.

47 Ibid., July 27, 1851; "Victor M. Wallace," B. F. Alley and J. P. Munro-Fraser, *History of Clarke County, Washington Territory Compiled from the Most Authentic Sources; also Biographical Sketches of its Pioneers and Prominent Citizens*, (Portland, Oregon: Washington Publishing Co., 1885), http://www.freepagesw.genealogy.rootsweb.com/~jtenlen/Clarke-history.html.

48 Butler Ives, July 28–29, 1851, MSS 3011.

49 Robert E. Ficken, *Washington Territory* (Pullman, Washington: Washington State University Press, 2002), 8.

50 Johansen and Gates, *Empire of the Columbia*, 134.

[51] Butler Ives, July 29, 1851, MSS 3011; Thomas Vaughan (ed.), "The Round Hand of George B. Roberts," *Oregon Historical Quarterly* 63 (1962): 102–107. A few months after he met Butler Ives, Roberts resigned his post, gained American citizenship, and took a donation land claim.

[52] BLM Field Notes, vol. OR-R0002.p. 0251.0, Township 8 North, Range 1 West (East Boundary); vol. OR-R0002, p. 0266.0, T 9 North, Range 1 West (East Boundary); vol. WA-R0002 p. 0047.0 Township 10 North, Range 1 West (East Boundary); vol. OR 0002. p. 0451.0, Township 4 North, Range 1 West (East Boundary). There are several places in Oregon and Washington named for the meridian including Meridian Mountain in Washington; Meridian, a town in Marion County, Oregon; Meridian Road in Olympia, Washington; and Meridian Roads in Jackson County, Woodburn, Silverton and Mount Angel, Oregon.

[53] Butler Ives, August 1, 1851, MSS 3011.

[54] Ibid., August 3, 1851.

[55] BLM Field Notes, vol. OR-R-0002, p. 0453.0 Township 4 North, Range 1 West (East Boundary); vol. WA-R0002 p. 0048.0–0049.0, Township 10 North, Range 1 West (East Boundary); WA-R0002 p. 0054.0 Township 11 North, Range 1 West (East Boundary).

[56] Ibid., August 4–6, 1851, MSS 3011. Active in territorial political affairs, Joseph Borst settled a few years earlier on Ford Prairie in the Chehalis River Valley.

[57] William Ives to John Preston, July 22, 1851; Series 5, vol. 1, RG 49, NARA-Seattle.

[58] Butler Ives, August 7–11, 1851, MSS 3011.

[59] BLM Field Notes, WA-R0002 p. 0083.0 Township 19 North, Range 1 West (East Boundary).

[60] William Ives, August 16, 1851, MSS 1126.

[61] BLM Field Notes, WA-R0002 pp. 0084.0–0085.0, Township 19 North, Range 1 West (East Boundary).

[62] Ibid.

[63] William Ives to John Preston, September 6, 1851; Series 5, vol. 1, RG 49, NARA-Seattle.

[64] Ibid.

Chapter 4

[1] David Alan Johnson, *Founding the Far West: California, Oregon, and Nevada, 1840–1890* (Berkeley, California: University of California Press, 1992), 53.

[2] Asahel Bush to John Preston, July 21, 1851; Series 3, vol. 1, p. 106, RG 49, NARA-Seattle.

[3] Jeff LaLande, "'Dixie' of the Pacific Northwest: Southern Oregon's Civil War," *Oregon Historical Quarterly* 100, no. 1 (1999): 36.

[4] Ibid., 37–38.

[5] Dorothy O. Johansen, "The Roll of Land Laws in the Settlement of Oregon," *Genealogical Material in Oregon Donation Land Claims*, vol. 1, Genealogical Forum of Portland, Oregon, Portland: 1957, n.p.

[6] Thomas Nelson to Cornelia Nelson, October 4, 1851, WA MSS. 2176.

[7] *Oregon Spectator*, August 15, 1851; *Portland Oregonian*, August 19, 1851 2:5.

[8] Preston planned these first contracts for Townships 1 through 9 South, Ranges 1 to 4 West and Range 1 East. Preston to Butterfield, October 20, 1851; Series 2, vol. A, p. 27, RG 49, NARA-Seattle; Norman Caldwell, e-mail to the author, July 22, 2002.

[9] Butler Ives, August 16, 1851, MSS 3011.

[10] Butler Ives, August 18–19, 1851, MSS 3011. Ives' diaries overlap between August 15–August 20, 1851; he apparently copied earlier diary entries into a fresh volume.

[11] Butler Ives, August 23; 25, 1851, August 15, 1851, MSS 3011; Butler Ives, Letters and two documents, Container 6, Folder 1, Papers of M. A. Ives, 1850–1872, Stanford University Libraries, Department of Special Collections and Archives.

[12] White, "The Public Land Surveys in Oregon."

[13] Butler Ives, August 26, 1851, MSS 3011.

[14] *Oregon Spectator*, August 19, 1851 2:2.

[15] Butler Ives, August 26–27, 1851, MSS 3011.

[16] Minnick, 266–267.

[17] Ibid., 259–260.

[18] Butler Ives, August 28–September 6, 1851; Prettyman's donation land claim was located in Townships 1 South, Range 1 East, and 1 South, Range 2 East; Grant Nelson, "The Early Years of Mt. Tabor," Mt. Tabor Neighborhood Association Newsletter (n.d) http://www.mttaborpdx.org/history_early_years.html; Lewis Love settled in 1850 in Township 1 North, Range 1 East. Albert Kelly's claim was in Township 1 South, Range 1 East.

[19] Butler Ives, September 9, 1851, MSS 3011.

[20] Ibid., September 15, 1851.

[21] BLM Field Notes vol. OR-R0005 p. 0249.0–0251.0, Township 1 North, Range 1 East (General Description).

[22] BLM Field Notes, vol. OR-R0005 p. 0253.0; Township 1 North, Range 1 East; vol. OR-R0003 p. 157.0, Township 1 South, Range 1 East; Butler Ives, September 22, 1851, MSS 3011; McFall was born January 23, 1829 in Trumbull County, Ohio to Henry and Mary Haney McFall. The other new crew members were J. J. Cook, John H. Jones, and J. Franklin Keen.

[23] McFall, *Oregon Journal*, September 25, 1851, typescript held in the office of the Jackson County Surveyor, Medford, Oregon; October 2, 1851; Butler Ives, October 21, 1851, MSS 3011.

[24] Contract No. 4 called for the survey of the first standard parallel south of the base line for twelve miles east of Willamette Meridian and for twenty-four miles west, also the lines for Townships 6, 7 and 8 South, Ranges 2, 3, and 4 West. Ives hired former crew members Clark and Lambert as chainmen and Swinhart as axeman; Preston to Butterfield, September 24, 1851; September 27, 1851; Series 2, vol. A, pp. 14–16; RG. 49; NARA-Seattle.

[25] William Ives to John Preston, September 29, 1851; October 5, 1851, Series 5, vol. 1 RG 49; NARA-Seattle.

[26] Preston to Butterfield, September 24, 1851; Series 2, vol. A, p. 14, RG 49, NARA-Seattle. Preston wrote Commissioner Butterfield, "In settling with Mr. Freeman, I paid him for the lines he run west (24 miles in all) at the rate of $18 per mile, the same as the township lines — for all other lines (north & south ones) the same as the Meridian $20 per mile." Contract No. 5 called for the survey of Townships 6, 7, and 8 South, Range 1 West and Townships 9 and 10 South, Ranges 1, 2, 3, and 4 West; Freeman's contract also included the survey of the Second Standard Parallel between Townships 10 and 11 South through Ranges 1, 2, 3, and 4 West of the Willamette Meridian.

[27] George Hyde to John Preston, October 20, 1851; Series 5, vol. 1, item 38, RG 49, NARA-Seattle; *Oregon Spectator*, August 15, 1851 2:2; September 8, 1851, 2:5; December 9, 1851, 3:2. Moody and Seymour had each assisted the deputy surveyor on the meridian survey. In August, 1851, Allan Millar was named a clerk of the Territory's supreme court.

[28] Oregon City *Oregon Statesman*, September 2, 1851, 3:2.

[29] Webster, *Gold Seekers of '49*, 196.

[30] Ibid., 203–204.

[31] Ibid., 204–205.

[32] Applegate to Preston, September 19, 1851; Series 3, vol. 1, item 34; RG-49; NARA-Seattle.

[33] William Ives to John Preston, September 9, 1851; Series 5, vol. 1, item 6, RG 49, NARA-Seattle. Hunt was also recommended by Charles Noble, Michigan Surveyor General, March 14, 1851; Series 5, vol. 1, p. item, RG 49, NARA-Seattle; Genealogical records appear to confirm that Daniel Adams Thurston, born July 16, 1808 in New Hampshire was Samuel R. Thurston's brother.

[34] Howard McKinley Corning, *Dictionary of Oregon History* (Portland, Oregon: Binfords & Mort, 1956), 70; BLM Field Notes, vol. OR-R0004 p. 0358.0; vol. OR-R0004 p. 0696.0, Township 8 South, Range 1 West (Subdivisions); Robert Paxton, Melvin L. Hutton, Allan M. Seymour, and Thomas O. Davis also assisted Hyde on this contract.

[35] Contract 6, Special Instructions; Series 22, Contracts and Bonds 1851–1870; RG 49 NARA-Seattle; Minnick, 270–272; Norman Caldwell, e-mail to author, July 22, 2002.

[36] Minnick, 261. Hyde subdivided Townships 6 and 7 South, Range 1 West.

[37] Boag, *Environment and Experience*, 52.

[38] Ibid., 117.

[39] Butler Ives, October 24–29, 1851; Johansen and Gates, *Empire of the Columbia*, 231; Robert Elder surveyed the Oregon City Lots; Butler Ives surveyed McLoughlin's donation land claim boundaries; Butler Ives, October 28 1851, MSS 3011; Contract 8 encompassed subdivisions of Townships 1 North and Townships 1, 2, 3, 4, and 5 South, Range 1 East of the Willamette Meridian.

[40] William Ives' Contract No. 9 included Townships 6 and 7 South, Ranges 3 and 4 West and Township 8 South, Range 3 West; Contract No. 10 included Township 1 North, Ranges 2 and 3 West; Freeman's Contract 11 was for Township 8 South, Range 2 West, Township 9 South, Ranges 1, 2, and 3 West, and Township 10 South, Range 1 West.

[41] White, "The Public Land Surveys in Oregon"; White notes that after the plat and field notes were in order and approved by the Surveyor General, copies were sent to the Commissioner of G.L.O. in Washington, DC while the originals remained with the Surveyor General.

[42] Rod Squires, "Cartographic Records of the Public Land Survey in Minnesota."

[43] Preston to Butterfield, September 1, 1851; Series 2, vol. A, p. 9, RG 49, NARA-Seattle.

[44] Ibid.

[45] Ibid., September 1, 1851; September 27, 1851, 9;14.

[46] Jeff LaLande, personal communication, November 30, 2004.

[47] Oregon State Archives, *Guide to Oregon Provisional and Territorial Government Records*, (Salem, Oregon: State Printing Division, 1990), 80–82.

[48] *Oregon Statesman*, December 23, 1852, 1:2.

[49] Ibid. The *Spectator's* article ran on August 15, 1851, and the *Portland Oregonian* printed a similar article on August 19, 1851, 2:5.

[50] *Oregon Statesman*, December 23, 1851, 1:2.

[51] Ibid.

[52] Ibid.

[53] Ibid.

[54] George McFall, December 18–19, 1851.

[55] William Ives to John Preston, December 20, 1851; Series 5, vol. 1, RG 49, NARA-Seattle. William Ives' assistants on this contract were J. M. Clark, James Dallas Price, James

Swinhart, and Zachariah Dotson. Illinois native James Dallas Price was born in 1830. He eventually settled a land claim in Linn County, Township 11 South, Range 3 West. Zachariah S. Dotson acquired land through sale cash entry in Clackamas County.

[56] George McFall, December 25, 1851.

Chapter 5

[1] Webster, *The Gold Seekers of '49*, 203.

[2] *Oregon Spectator*, February 2, 1852, 2:2; BLM Field Notes, vol. OR-R0007 p. 0251.0 (Subdivisions, Township 8 South, Range 3 West, Section 14 and 15); William Ives, February 2, 1852, MSS 1126.

[3] Evans to Preston, [Spring] 1852; Series 3, vol. 1, 206, RG 49, NARA-Seattle. Ives found the fossils in Township 6 South, Range 4 West.

[4] William A. Burt, A *Key to the Solar Compass, and Surveyor's Companion*, (Philadelphia, Pennsylvania: William S. Young), 1855, 76.

[5] BLM Field Notes, vol. OR-R0008 p. 0521.0, Township 9 South, Range 3 West (Subdivision).

[6] Webster, *The Gold Seekers of '49*, 203.

[7] Ibid., 201–202.

[8] Ibid.

[9] Freeman to Preston, March 4, 1852; Series 5, vol. 1, Item 8, RG 49, NARA-Seattle.

[10] William G. Robbins, *Landscapes of Promise: The Oregon Story 1899–1940* (Seattle, Washington: University of Washington Press, 1997), 84.

[11] BLM Field Notes, vol. OR-R0008 pp. 0304.0–0305.0, Township 9 South, Range 1 West (Subdivision).

[12] Boag, *Environment and Experience*, 65.

[13] Ibid., 63.

[14] Freeman to Preston, March 16, 1852; Series 5, vol. 1, Item 9, RG 49, NARA-Seattle.

[15] James A. Cardwell, "Emigrant Company," MSS P-A, University of California, Bancroft Library, Berkeley, California.

[16] George Hyde to John Preston, February 10, 1852; Series 5, vol. 1, Item 7, RG 49, NARA-Seattle. Hyde had finished subdividing Township 6 South, Range 1 West.

[17] White, *A Casebook of Oregon Donation Land Claims*, 22. The announcement is dated February 5, 1852; it was published on February 10, 1852, in the *Oregon Spectator* and at approximately the same time in the *Oregon Statesman* and *Portland Oregonian* newspapers.

[18] *Oregon Spectator,* March 16, 1852, 3:2.

[19] Lane J. Bouman, "The Location and Survey of Oregon Donation Land Claims," Part I, *The Oregon Surveyor*, 21, no. 4 (1999):14–15.

[20] Hiram Colver to John Preston, May 24, 1853; Series 3, vol. 1, item 7, RG 49, NARA-Seattle.

[21] George Hyde to John Preston, February 10, 1852; Series 5, vol. 1, item 7, RG 49, NARA-Seattle. Hyde wrote from Township 6 South, Range 2 West.

[22] J. F. Keen, Zachariah Dotson, and William A. Potter also assisted Ives. Potter, born in Trumbull County, Ohio, in 1825, arrived in Oregon in the fall of 1851. Joseph Gaston, *Centennial History of Oregon, 1811–1912*, vol. 2 (Chicago, Illinois: S. J. Clarke Publishing Company, 1912), 67; Butler Ives, March 1, 1852, MSS 3011. William McDermott and William Doak also worked at various times on this contract. Ives resumed work in Township 5 South, Range 1 East.

[23] Butler Ives to John Preston, March 5, 1852; Series 5, vol. 1, item 46, RG 49, NARA-Seattle.

[24] McFall, March 1, 1851; Butler Ives, March 5, 1852, MSS 3011. The Gribbles were married November 9, 1851, in Clackamas County. Andrew Gribble died April 19, 1879, and is buried in the Gribble Cemetery in Clackamas County. Nancy Gribble died April 9, 1925.

[25] Butler Ives, MSS 3011, Folder 2; William Ives, March 17, 1852; August 27, 1852, MSS 3011, Folder 3.

[26] Butler Ives, March 27, 1852, MSS 3011.

[27] BLM Field Notes, vol. OR-R0005 p. 0391.0–0392.0 (Township 1 South, Range 1 East (Subdivision).

[28] BLM Field Notes, vol. OR-R0006 pp. 0241.0–0242.0, Township 3 South, Range 1 East (Subdivision).

[29] Webster, *The Gold Seekers of '49*, 206. For Contract 15, Elder surveyed boundaries for Townships 1 through 5 South, Ranges 3 and 4 West.

[30] Webster, *The Gold Seekers of '49*, 207; The Murphys were sons of Marion County donation land claimants Daniel and Catherine Dillon Murphy; Preston issued Contract 20 on June 8, 1852. Estimated at 660 miles, it called for the subdivisions of Townships 3 and 5 South, Range 2 West; Townships 1, 3, 4, and 5 South, Range 3 West, and Townships 2, 3, 4, and 5 South, Range 4 West.

[31] BLM Field Notes, vol. OR-R0011 p. 0081.0, Township 8 South, Range 4 West (Subdivision). Freeman's contract included the survey of Townships 8 and 9 South, Range 4 West and Township 10 South, Ranges 2, 3, and 4 West. In addition to Zenas Moody, crew members included George Plummer and Daniel A. Thurston, chainmen; Abram Ensley was the axeman. George R. Lee and Noah G. Herren assisted in Township 10 South, Range 2 West and in Township 8 South, Range 4 West. Henry Hilton worked with George Plummer as a chainman.

[32] BLM Field Notes, vol. OR-R0011 p. 0382.0, Township 10 South, Range 3 West (Subdivision).

[33] BLM Field Notes, vol. OR-R0011 p. 0161, Township 9 South, Range 4 West (Subdivision).

[34] William Ives, April 10, 1852, MSS 3011; John Alva Harry, a native of Indiana, had arrived in Oregon the previous October. Isaiah Case and his brother, William, settled in the Champoeg vicinity in the early 1840s. Twenty-year-old Sylvester Cannon was a native of Ohio. Contract 12 called for the survey of Townships 1–5 South, Ranges 1–2 West.

[35] Philip Dole, Professor of Architecture Emeritus, University of Oregon, note to the author, August 24, 2005.

[36] *Oregon Spectator*, December 9, 1847, 2:1; Corning, 175.

[37] John A. Hussey, *Champoeg: Place of Transition, A Disputed History* (Portland, Oregon: Oregon Historical Society Press, 1967), 195, 197, 201, 208–209.

[38] William Ives, MSS 3011; James V. Walker, M.D., "Glimpses of Pioneer Life: The Survey Plat and the Government Land Office in Oregon," *Exploring Mercator's World*, vol. 7 No. 2 (March–April 2002).

[39] William Ives, MSS 3011. Ives' expenses included four drinks on May 18, a quart of whiskey on June 6, and a half-pint of whiskey on July 28.

[40] Contract 18 was for subdivisions of Townships 1, 3, 4, and 5 South, Range 1 West and Township 4 South, Range 2 West; William Ives, May 21, 1852, MSS 3011.

[41] John Harry went to Salem for supplies on June 9 and on June 18. James D. Price bought provisions at Butteville on June 25 and in Oregon City on July 22. William Ives and his men went to Oregon City on July 28 to make their oaths and buy more provisions.

[42] William Ives, June 19, 1852, MSS 3011. Ibid., July 8, 1852; July 9, 1852; July 30, 1852.

[43] William Ives, March 21; 28, 1852, MSS 3011; Butler Ives, May 25, 1851; August 24, 1851, MSS 3011.

44 William Ives, June 12–13, 1852, MSS 3011. Samuel Allen secured his land claim on Abiqua Creek in Township 6 South, Range 1 West, Sections 13, 14, 23, and 24 in March, 1847.

45 *Portland Oregonian* June 12, 1852 3:3; William Ives, June 6–7, 1852, MSS 3011. Judy Card, "Early Families in the Woodland Community: A 1958 Report of the Woodland History Committee," http://www.lewisriver.com/pt2-page2.html (February 23, 2004).

46 William Ives, June 18, 1852, MSS 3011.

47 Butler Ives to William Ives, May 3, 1862, Timothy Hopkins Transportation Collection, (M097, Box 6, Folder 1), Department of Special Collections and University Archives, Stanford University Libraries, Stanford, California.

48 Loren Williams to William Ives, September 25, 1851; Series 3, vol. 1, p. 140, RG 49, NARA-Seattle; Beckham, *Land of the Umpqua*, 76;90.

49 William Ives, July 3, 1852, MSS 3011.

50 The clerks chided the deputies for sloppy work. Wells Lake requested that the surveyor more carefully enter his field notes in separate, orderly lines, adding, "In consequence of the large number of erasures I would recommend Mr. Freeman to get a bottle of pounce." BLM Field Notes, vol. OR-R0003, p. 0.450.0, Township 10 South, Range 4 West (Subdivisions). Pounce is a powder used to prevent ink from spreading on the page.

51 N. Dubois, a surveyor and draftsman, arrived in Oregon in 1851. He worked as a surveyor at Milwaukie, Oregon, until the spring of 1852 when he began work in Preston's office. Little is known about DuBois, including his first name. He signed only as "N. DuBois." He is listed as a draftsman for the 1868 "United States Map of the Territory" (William J. Keeler). Civil engineer George H. Belden worked for several years in the surveyor general's office before resigning to work for the Oregon California Railway Survey; George Belden to John Preston, April 6, 1852; Series 3, vol. 1, p. 33, RG 49, NARA-Seattle.

52 Oregon City *Oregon Spectator,* August 19, 1853, 1:2. Son of Eli and Ruth Hurd Lake, Wells Lake had known the Prestons in Illinois; Wells Lake to John Preston, December 2, 1851; Series 3, vol. 1, item 139, RG 49, NARA-Seattle. Lake wrote, "Remember me to Mrs. Preston, also to Mr. Robt. Elder who I presume is with you."

53 William Ives, MSS 3011. These figures are part of the expense account at the back of the volume.

54 On Contract 21, Ives surveyed in Township 2 South, Range 1 West, and Township 1 South, Range 2 West.

55 Butler Ives, August 19–26, 1852, MSS 3011; McFall, August 23, 1852. The reserve was formally abandoned in 1869 by order of the Secretary of War.

Chapter 6

1 William Ives, April 1, 1852, MSS 3011; William Ives to Butler Ives, January 3, 1864, MSS 1126; William Ives, August 30, 1852, MSS 3011.

2 William Ives, August 27, 1852, MSS 3011; Butler Ives, August 21, 1852 (Expense Account); August 30, 1851, MSS 3011.

3 William Ives, August 31–September 2, 1852, MSS 3011.

4 Butler Ives, August 31, 1852; September 2–4, 1852, MSS 3011. Contract 22 called for the survey of boundaries for Townships 1–10 South, Range 5 West.

5 Martindale, born in Indiana in 1829, came west in 1849. He later settled a donation land claim in Camas Valley, Douglas County, Oregon; Butler Ives, September 6; 13, 1852, MSS 3011.

[6] Butler Ives, September 26–27; October 2, 1852, MSS 3011; McFall, September 26, 1852.

[7] Butler Ives, October 14, 1851, MSS 3011; Johansen and Gates, *Empire of the Columbia*, 135. Several families established the nucleus of a Canadian population in this area and named their settlement French Prairie.

[8] Butler Ives, October 25, 1851, MSS 3011.

[9] William Ives, September 9, 1852, MSS 3011.

[10] Ibid., September 12, 1852. William Ives' description of Jacksonville is one of the earliest known accounts of the town's appearance.

[11] Ibid.

[12] Ibid., September 13, 1852.

[13] Ibid. Ives went by stage from Shasta City to Sacramento, arriving there on September 23, 1852. After visiting the gold country near Grass Valley, he went to San Francisco. Ives left for Panama on October 15, 1852, and reached New Orleans on November 12 and Detroit on November 29, 1852.

[14] White, *A Casebook of Oregon Donation Land Claims*, 23, 27. Jammed with requests, the General Land Office lagged behind in issuing patents; Lane J. Bouman, "The Location and Survey of Oregon Donation Land Claims," Part I, "The Role of the Surveyor in the Initial Location of Donation Land Claims" *The Oregon Surveyor* 31, no. 4 (1999): 15.

[15] Surveyor General Preston issued the first two contracts on August 21 and the third on September 11, 1851; White, "The Public Land Surveys in Oregon."

[16] White, *A Casebook of Oregon Donation Land Claims*, 115.

[17] Green surveyed Townships 6, 7, and 8 South, Ranges 1 and 2 West; Israel Mitchell, Township 1 South, Ranges 2 and 3 West.

[18] White, *A Casebook of Oregon Donation Land Claims*, 23.

[19] Beckham, *Land of the Umpqua,* 162.

[20] Butler Ives to William Ives, December 2, 1852, private collection; Johnson, *Founding the Far West*, 43.

[21] White, *A Casebook of Oregon Donation Land Claims*, 21.

[22] Letters from DLC Claimants A–C, MSS 914.

[23] *Oregon Statesman*, March 26, 1853, 2:2. A check of the calculations reveals a discrepancy in the total figures of acreage and numbers of settlers.

[24] Ibid.

[25] *The Columbian* began publication on September 11, 1852.

[26] Fred Yonce, "Public Land Surveys in Washington" *Pacific Northwest Quarterly* 63, no.4 (1972): 130–131.

[27] Webster, *The Gold Seekers of '49*, 208–209. Kimball Webster's first solo contract was dated December 1, 1852.

[28] Ibid., 209.

[29] Butler Ives to William Ives, December 2, 1852, private collection; Ives signed Contract 28 on December 1 and began work on December 23, 1852.

[30] Preston to Butterfield, November 24, 1852; Series 2, vol. A, p. 128, RG 49, NARA-Seattle.

[31] Freeman to Preston, [July] 1852, Series 5, vol. 1, item 22, RG 49, NARA-Seattle. Contract 24 was for boundaries for Townships 14 South, Ranges 4 and 5 West; Township 15 South, Ranges 3, 4, and 5 West; and Townships 16, 17, and 18 South, Ranges 3 and 4 West. Crew members Plummer, Chittenden, and Moody assisted Freeman; others working included Dennis Hathorn, William Gilcrist, William Dobbins, and A. W. Stannard.

[32] Butler Ives to William Ives, December 2, 1852, private collection. Contract 25 called for subdividing Townships 1 through 5 South, Range 5 West.

[33] Butler Ives to William Ives, December 2, 1852, private collection.

[34] Johansen and Gates, *Empire of the Columbia*, 242; *Oregon Statesman*, November 27, 1852 3:1. Governor Gaines served from August 18, 1850 to May 16, 1853. After the death of his first wife, he married Margaret Wands, one of the teachers escorted by the Prestons to Oregon City.

[35] *Portland Oregonian*, December 18, 1852, 2:2. The *Charles Devens* was among the fastest sailing vessels of the time. As territorial residents, Oregonians were not eligible to vote in this election.

Chapter 7

[1] Butler Ives to William Ives, February 4, 1853, Timothy Hopkins Transportation Collection.

[2] *Portland Oregonian,* January 8, 1853, 2:3.

[3] Butler Ives to William Ives, February 4, 1853.

[4] Webster, *The Gold Seekers of '49*, 210, 212.

[5] Ibid., 210–211, 213. Webster mistakenly wrote the uncooperative employee's name as McDonald; he was David McLellan. McClellan's companion was D. J. Homes.

[6] Webster, 214.

[7] McFall, December 18, 1853. In addition to McFall, crew members for this contract included John Richards, J. F. Keen, John Lewis, and Allen Potter. Blair's donation land claim lay in sections 25, 26, 35, and 36 of Township 17 South, Range 4 West. Blair is credited with providing lumber for and building the first Lane County Courthouse.

[8] McFall, December 25, 1852.

[9] Butler Ives to William Ives, February 4, 1853, Timothy Hopkins Transportation Collection.

[10] Ibid.

[11] *Portland Oregonian*, January 8, 1853, 2:6; Butler Ives, Account Book, MSS 3011.

[12] Butler Ives to William Ives, February 4, 1853, Timothy Hopkins Transportation Collection.

[13] Ibid.; Butler Ives refers to Orville C. Pratt, a Territorial Supreme Court judge. In *Eden Seekers, the Settlement of Oregon, 1848–1862*, Malcolm Clark, Jr. writes that during the summer of 1852 Pratt went east to lobby for the position of chief justice for the Territory.

[14] Butler Ives to William Ives, February 4, 1853, Timothy Hopkins Transportation Collection.

[15] John B. Preston to John S. Wilson, March 14, 1853; Series 2, vol. A, p. 137, RG 49, NARA-Seattle. Wilson succeeded Butterfield on September 16, 1852.

[16] White, *A History of the Rectangular Survey System*, 116.

[17] Johnson, *Founding the Far West*, 43.

[18] Johansen and Gates, *Empire of the Columbia*, 247–248; White, *A History of the Rectangular Survey System*, 117; Oscar Osburn Winther, "Inland Transportation and Communication in Washington 1844–1859," *Pacific Northwest Quarterly* 30 (1939): 371. The territorial census of 1853 counted 3,965 settlers living north of the Columbia River.

[19] *Oregon Statesman*, January 22, 1853, 1:4; January 29, 1853, 2:7.

[20] Elijah Dodson to John B. Preston, Letters from Donation Land Claimants, 1853. MSS 914, Oregon Historical Society Research Library, Portland, Oregon. Elijah Dodson held a 629-acre donation land claim in Yamhill County. His second wife, Elizabeth Hines Dodson, died in Oregon in 1854.

[21] Preston W. Gillett to John B. Preston, MSS 914.

[22] Lewis Cline to John B. Preston, MSS 914.

[23] Ellen Smith to John Preston, January 18, 1853, MSS 914.

[24] Elder's contract No. 31 was dated April 7, 1853. He shared this contract for township boundaries and subdivisions with Henry Gile; Ives' contract, No. 32, was dated April 9, 1853; McFall, April 18, 1853. These townships encompassed 9 through 12 North, Ranges 1 and 2 West.

[25] George Hyde's and Josiah Preston's contract called for surveying the meridian through Townships 20 and 21 North into Puget Sound, as well as several townships. Preston gave Joseph Latshaw Contract 34 for subdivisions of four townships in the vicinity of Eugene City. Hyde's and Latshaw's contracts bore the date of April 16, 1853. For Contract 34, Latshaw subdivided Townships 15–18 South, Range 3 West.

[26] "Members of the Legislature of the State of Oregon and Chief Clerks, 1843–1967", Oregon State Archives, Bulletin No. 2 Revised, Publication No. 30, 1968; White, *History of the Rectangular Survey System*, 222; White, "Harvey Gordon: US Deputy Surveyor and Designer of the Oregon State Seal," *The Oregon Surveyor* 35, no. 2 (2002): 18–22. Gordon, who became chief clerk in the Surveyor General's Office in 1856, designed Oregon's state seal in 1857. Elected state printer in June, 1862, he died the following month of tuberculosis.

[27] John Blaine Kerr, *Biographical Dictionary of Well-known British Columbians: With a Historical Sketch* (Vancouver, British Columbia: Kerr & Begg, 1890), 311; Herbert O. Lang (ed), *History of the Willamette Valley* (Portland, Oregon: Himes and Lang, 1885), 622. Ford accompanied a party in 1846, checking the route from the Willamette Valley to the Snake River and, in 1847, explored the Rogue and Umpqua country. Vanderpool's first name is occasionally spelled Meadors or Meadows.

[28] White, *A History of the Rectangular Survey System*, 117, 199. Hays was appointed California Surveyor General on March 19, 1853; Freeman to Preston, May 9, 1853, Series 3, Box 50, vol. 4, p. 100, RG 49, NARA-Seattle. Freeman wrote Preston from Albany, Oregon, to recommend a man for a surveying position. He apparently left Oregon in late May or early June, 1853; Thomas W. Prosch, "Notes on Oregon Conditions in the Fifties," *Oregon Historical Quarterly* 8, no. 2 (1907): 191–200. Freeman joined his youngest brother, Jonathan, in San Francisco; Chautauqua County Probate Record, vol. 1, p. 206, Fenton History Center Museum & Library, Jamestown, New York. Jonathan Freeman, James Freeman's father, wrote his will on December 17, 1853. This document declares that both his sons were residents of San Francisco.

[29] McFall, April 20, 1853. The crew included George McFall, John N. Lewis, Squire Moon, J. F. Keen, Frank Hopkins, and William B. French.

[30] McFall, June 1–4, 1853.

[31] Ibid., August 24, 1853.

[32] Maria Cable Cutting, "After Thoughts," *Oregon Historical Quarterly* 63 (1962), 240–241. The men in the company were R. M. Brandenburg, R. J. Dixon, S. R. Riffle, and William Byers. The Cutting's farm was located in Township 12 North, Range 2 West, Section 2.

[33] BLM Field Notes, vol. WA-R00020, p. 0456.0. Their crew consisted of Timothy W. Davenport, J. M. Martin, E. L. Merrill, C. C. Lytle, Robert Cummins, Sylvester Cannon, and B. R. Hanley.

[34] Josiah W. Preston to John B. Preston, September 18, 1853, MSS 914. As part of their contract, George Hyde and Josiah Preston also surveyed two standard parallels. The contract called for surveying Townships 17, 18, and 19 North, Range 1 West; Townships 17, 18, and 19

North, Range 2 West, Townships 17, 18, and 19 North, Range 1 East, as well as the exterior lines for Townships 20 and 21 North, Ranges 1 East and 1 West. The surveyors completed the contract on October 21, 1853.

35 *Oregon Statesman*, April 30, 1853, 2:7.

36 Ibid., April 23, 1853, 2:6; April 30, 1853, 2:3.

37 Ibid., May 21, 1853, 1:1.

38 http://arcweb.sos.state.or.us/banners/governors.htm; *Oregon Statesman*, April 23, 1853, 2:7.

39 *Oregon Statesman*, February 26, 1853 2:1.

40 Louis Hyde Family Book, Reed Hyde Papers. George Hyde also invested in the business. James O'Neill should not be confused with another settler of the same name; Jewel Lansing, *Portland: People, Politics, and Power, 1851–2001* (Corvallis, Oregon: Oregon State University Press, 2003), 83; *Oregon Statesman*, April 30, 1853, advertisement; *Oregon Statesman*, April 30, 1853, 2:7.

41 Hubert Howe Bancroft, *History of Oregon*, vol. II (San Francisco, California: The History Company, 1888), 159; "Reminiscences of Mrs. E. W. Wilson," W13909, WPA Life Stories, http://www.rootsweb.com/~ormultno/Stories/wpa/Wilson.htm.

Chapter 8

1 *Senate Executive Journal*, Monday February 7, 1853, "A Century of Lawmaking for a New Nation: US Congressional Documents and Debates, 1774–1875," American Memory Collection, http://memory.loc.gov/ammem/mdbquery.html. Shortly before leaving office, President Fillmore nominated Allan P. Millar as US Marshall for Oregon Territory. When Franklin Pierce assumed office in March, 1853, he named J. W. Nesmith to the position instead. Ralph Wilcox (1818–1877) came to Oregon in 1845. He served in the Provisional Legislature as Speaker of the House for the Territorial Legislature in 1850–1851, and as president of the council in 1853–1854.

2 *Portland Oregonian*, Portland, June 25, 1853 2:2.

3 *Oregon Statesman*, July 5, 1853, 2:6. The *Oregon Statesman* began publication in Salem in June, 1853.

4 White, *A History of the Rectangular Survey System*, 219.

5 *Oregon Spectator*, August 19, 1853, 2:3.

6 Dennis Hathorn to John Preston, MSS 914.

7 *Oregon Spectator*, August 19, 1853, 2:3; September 16, 1853, 4:2; *Oregon Statesman*, August 16, 1853, 2:7.

8 John Preston to John Wilson, September 1, 1853; Series 3; vol. 4, p. 154, RG 49, NARA-Seattle. *Senate Executive Journal*, Monday, January 16, 1854, "A Century of Lawmaking for a New Nation: US Congressional Documents and Debates, 1774–1875," American Memory Collection, http://memory.loc.gov/ammem/mdbquery.html. Franklin Pierce nominated Charles K. Gardner of the District of Columbia during the Senate recess.

9 Thomas McF. Patton to John Preston, November 1, 1853; Series 3, vol. R, item 195, RG 49, NARA-Seattle.

10 *Oregon Spectator*, August 19, 1853 to January 7, 1854 (advertisement); *Oregon Spectator*, August 19, 1953, 2:2, McLoughlin completed two large buildings on Main Street; Preston and O'Neill occupied one; "Costs of Improvements Made by Dr. John McLoughlin," *Oregon Historical Quarterly*, 14 (1913): 70.

11 Contract 31 was for Townships 23 through 28 South, Ranges 5 and 6 West; Beckham, *Land of the Umpqua*, 76. Henry Gile assisted Elder at surveying for this contract.

[12] Kimball Webster, *The Gold Seekers of '49*, 214–215.

[13] Applegate to Preston, Surveyor General, August 21, 1853; Series 3, vol. 4, p. 15, RG 49, NARA-Seattle; Webster, 215. Elder returned to his farm in Kankakee, Illinois, in late 1853. He moved with his brothers to Olmsted County, Minnesota, in the 1860s and worked there as a civil engineer. Elder died after 1870.

[14] John Preston to John Wilson, October 13, 1853; Series 3, vol. 4, p. 177, RG 49, NARA-Seattle.

[15] Applegate to Preston, September 2, 1853; Series 3, vol. 4, p. 174, RG 49, NARA-Seattle.

[16] Ibid.

[17] Applegate to Preston, September 2, 1853; Series 3, vol. 4, p. 133, RG 49, NARA-Seattle.

[18] Beckham, *Land of the Umpqua*, 155; *Oregon Statesman*, September 27, 1853, 2:2.

[19] Anson Henry, appointed surveyor general of Washington Territory in 1861, died in 1865 when the steamer *Brother Jonathan* sank off the Oregon Coast.

[20] Webster, *The Gold Seekers of '49*, 219, 220. Kimball Webster left Oregon in 1854. Stopping for a few years in Missouri, he moved east where he worked as a surveyor in Hudson, New Hampshire. He married Abiah Cutter in 1857. Webster died June 29, 1916, at the age of 87 years.

[21] BLM Field Notes, vol. OR-R0007, p. 0092.0; OR-R0027, p. 675.0.

[22] McFall, November 21, 1853.

[23] *Oregon Statesman*, November 8, 1853, 2:5; *Oregon Spectator*, November 5, 1853, 2:3; *Portland Oregonian*, November 5, 1853, 2:6.

[24] Charles K. Gardner biography, http://www.famousamericans.net/charleskgardner; http://www.congressionalcemetery.org; Benson J. Lossing, *The Pictorial Field-Book of the War of 1812 or Illustrations, by Pen and Pencil, of the History, Biography, Scenery, Relics, and Traditions of the Last War for American Independence* (New York, New York: Harper & Brothers Publishers, 1868), 804.

[25] *Oregon Statesman*, November 15, 1853: 2:1.

[26] Webster, *The Gold Seekers of '49*, 222.

[27] *Oregon Spectator*, November 26, 1853 2:6; *Oregon Statesman*, November 29, 1853, 1:7; Rockwood, ed., "Diary of G. H. Atkinson 1847–1858," 222.

[28] *Oregon Statesman*, September 9, 1851 2:4; *Oregon Spectator*, September 9, 1851, 2:5. Preston was admitted to the Oregon Bar, US District Court for Clackamas County, in 1851; Gardner to Preston, March 6, 1851; Series 3, vol. 2, p. 56, RG 49, NARA-Seattle; *Oregon Statesman*, April 11, 1854, 4:4 (advertisement); *Oregon Spectator*, December 24, 1853, advertisement; January 7, 1854, 2:6.

[29] *Oregon Statesman*, May 6, 1856 2: 2–4.

[30] Applegate to Gardner, November 12, 1853; Series 3, vol. 4, p. 220, RG 49, NARA-Seattle.

[31] Abraham Smith to Charles K. Gardner, November 20, 1853; Series 3, vol. 4, p. 222, RG 49, NARA-Seattle.

Chapter 9

[1] Gardner to Wilson, April 8, 1854, Series 2, vol. A, 1851–1859, p. 232, RG 49, (NARA-Seattle).

[2] Joseph Hunt, Probate File, Oregon State Archives, File No. A-44.

[3] Applegate to Gardner, January 25, 1854, Series 5, vol. 1, item 220, RG 49, (NARA-Seattle).

[4] Butler Ives, Letters and two documents, Papers of M. A. Ives. Ives' commission as deputy surveyor by C. K. Gardner is dated January 4, 1854.

[5] Cartee remained in Oregon for over twelve years, serving as engineer of construction in 1862 for the Oregon Steam Navigation Company railroad line along the Columbia River. He moved to Idaho in 1863 and was appointed surveyor general for the Territory in 1866. Cartee died in Boise, Idaho, on September 2, 1891.

[6] Gardner to Davis, January 15, 1854, Oregon State Archives, Provisional and Territorial Records, Microfilm Reel No. 72, Document 10753. President Franklin Pierce appointed Davis as governor on December 2, 1853.

[7] The contract called for the survey in southwest Oregon, one of the least known areas in the Territory.

[8] W. H. Smith to Charles Gardner, March 14, 1854, Series 3, vol. 6, RG 49, Seattle, Washington (NARA-Seattle).

[9] Ibid.

[10] *Oregon Statesman*, April 4, 1854, 2:3.

[11] Ibid. This issue of the *Oregon Statesman* announced that L. F. Grover was acting editor of the newspaper while editor Bush was "temporarily absent to the States."

[12] *Oregon Statesman,* April 4, 1854, 2:5.

[13] McFall, April 8–19, 1854.

[14] BLM Field Notes, vol. OR-R0032, p. 0402.0. Truax's first name appears in the historic record as Sewall or Sewell. He signed his oath for this contract as Sewall Truax. *Genealogical Material in Oregon Donation land Claims*, vol. 3, (Portland, Oregon: Genealogical Forum of Portland, 1959), 2.

[15] McFall, April 26–30; May 1–2, 1854.

[16] Butler Ives, "Linear Field Book of Willamette Meridian," Original Survey Field Books of Ives and Hyde, 1854, vol. 1, Photocopy in Jackson County Surveyor's Office, Medford, Oregon.

[17] Volume 31 of the Original Survey Field Books of Ives and Hyde, 1854, contains geodetic notes; McFall, May 3, 1854.

[18] McFall, May 4–9, 1854.

[19] The offset extended between Townships 32 and 33 South to the boundary between Ranges 5 and 6 West. Butler Ives, "Linear Field Book of Willamette Meridian," Original Survey Field Books of Ives and Hyde, 1854, vol. 1, 36.

[20] Ibid., 38; McFall, May 13, 1854. In January 1854, the legislature formally changed the name of Grave Creek to Leland Creek, but the old name remained in common use and is today the stream's official name.

[21] BLM Field Notes, vol. OR-R0032, pp. 0135.0–0139.0, Township 39 South, Range 1 West (East Boundary).

[22] BLM Field Notes, vol. OR-R0032, pp. 0135.0–0139.0, Township 39 South, Range 1 West (East Boundary).

[23] Atwood and Lang, 1995, 149–150. In other examples, what they called "S. pine" is sugar pine, "tamarack" is lodgepole pine, and "balm of Gilead" is black cottonwood. The surveyors referred to deerbrush and wild lilac both as "lilac" and buckbrush as "chapparal."

[24] The surveyors returned on an offset line between Townships 35 and 36 South, the eighth standard parallel south of the base line.

[25] The land claim was situated in Township 36 South, Range 2 West, Sections 27 and 28, Jackson County Deeds, vol. 1, 259–261.

[26] Butler Ives to Wells Lake, May 18, 1854; Series 5, vol. 1, p. 119, RG 49, NARA-Seattle. By an Act of January 12, 1854, the Territorial Legislature changed the name of the Rogue River to Gold River.

[27] Butler Ives, "Linear Field Book of Willamette Meridian," 76.

[28] McFall, May 17–18, 1854.

[29] Ibid., May 19, 1854.

[30] Ibid.

[31] BLM Field Notes, vol. OR-R0032, p.0080.0, Township 35 South, Range 2 West (South Boundary).

[32] BLM Field Notes, vol. OR-R0032, p.0085.0.

[33] Ibid. pp. 0086.0–0088.0; McFall, May 19, 1854.

[34] BLM Field Notes, vol. OR-R0032, p. 0096.0 Township 35 South, Range 1 West (South Boundary).

[35] BLM Field Notes, vol. OR-R0032, pp. 0098.0–0099.0.

[36] Ibid., p. 0109.0.

[37] Lewis A. McArthur, *Oregon Geographic Names* (Portland, Oregon: Oregon Historical Society Press, 1974), 506, 584.

[38] BLM Field Notes, vol. OR-R0032, p. 0138.0, Township 39 South, Range 1 West (East Boundary).

[39] McFall, May 23, 1854.

[40] Butler Ives, "Linear Field Book of Willamette Meridian," vol. 1, 91–93.

[41] Ibid., 95.

[42] McFall, May 24, 1854. Wagner Butte's elevation is 7,255 feet above sea level.

[43] McFall, May 25, 1854. The south boundary of Township 39 South, Range 1 East is part of the ninth standard parallel.

[44] Jeff LaLande, *An Environmental History of the Little Applegate River Watershed, Jackson County, Oregon*, USDA Forest Service, Rogue River National Forest, 1995, 50.

[45] BLM Field Notes, vol. OR-R0032, p. 0139.0, Township 39 South, Range 1 West (East Boundary).

[46] Butler Ives to Charles Gardner, June 10, 1854; Series 5, vol. 1, item 109, RG 49, NARA-Seattle.

[47] George McFall, May 26, 1854.

Chapter 10

[1] Wallace Stegner, "History Comes to the Plains," *American Heritage* 8 (1957): 109.

[2] McFall, May 27, 1854; BLM Field Notes, Volume OR R0032, p. 0143.0. In the late afternoon the men went into town and signed their oaths before C. S. Drew, clerk of the Jackson County Board of Commissioners.

[3] McFall, May 28, 1854.

[4] Ibid., May 29, 1854; Contract No. 39; Series 22, RG-49; NARA-Seattle. McFall was compassman, Shunk and Cowne chainmen, Lewis axeman, and Addington spademan. Truax served as Hyde's compassman, while Cannon, Griffiths, and Price were chainmen and Payne an axeman.

[5] A. G. Walling, *History of Southern Oregon: Comprising Jackson, Josephine, Douglas, Curry and Coos Counties (*Portland, Oregon: A. G. Walling, 1884), 300.

[6] McFall, June 12, 1854; Butler Ives to Charles Gardner, June 10, 1854, Series 5, vol. 1, item 109, RG 49, NARA-Seattle.

[7] Township 37 South, Range 2 West (Subdivision), Original Survey Field Books of Ives and Hyde, 1854, Book 3, 70–73; Hyde worked in Township 38 South, Range 1 West until

July 1, 1854. BLM Field Notes, vol. OR-R-033, p. 0388A, Township 38 South, Range 1 West (Subdivision).

[8] McFall, June 13, 1854; Ives and his men surveyed boundaries for Townships 34, 35, 36, 37, 38, and 39 South, Range 1 East; Townships 36 South, Range 2 East, and Township 39 South, Range 2 East. The Willamette Meridian extension lay along the east boundary of Townships 35 and 34 South, Range 1 West.

[9] McFall, July 9, 1854.

[10] The surveyors set the last post at the corner of Sections 13 and 24 in Township 41 South. BLM Field Notes, vol. OR-R0032, p. 0428.0, Township 40 South, Range 1 East; McFall, July 10, 1854.

[11] Francis Landrum, "A Major Monument: Oregon-California Boundary, *Oregon Historical Quarterly* 72 (1971): 10–12.

[12] BLM Field Notes, vol. OR-R0032, pp. 0432.0–0433.0.

[13] BLM Field Notes, vol. OR-R0032, pp. 0433.0–0434.0.

[14] McFall, July 12, 1854. During the week of July 14–19, the territorial census-taker included the survey party in the tally for Jackson County; Jackson County Territorial Census, Provisional and Territorial Documents, Reel 78, Document 12281, Oregon State Archives, Salem, Oregon. The surveyors worked in Townships 34 South, Range 1 West, and 35 South, Ranges 1 and 2 West.

[15] Butler Ives, "Linear Field Book of Willamette Meridian;" McFall, July 19; 28–29, 1854.

[16] BLM Field Notes, vol. OR-R0033, p. 0235.0 (Subdivisions).

[17] BLM Field Notes, vol. OR-R0031 p. 0390.0.

[18] *Portland Oregonian*, August 5, 1854, quoted in Johansen and Gates, *Empire of the Columbia*, 243.

[19] *Oregon Spectator*, June 2, 1854, 3:4; Gardner to Wilson, August 1, 1854, Series 2, vol. A., p. 238, RG 49, NARA-Seattle.

[20] Gardner to Wilson, October 23, 1854; Series 2, vol. 1, p. 247, RG 49, NARA-Seattle.

[21] Butler Ives to Charles Gardner, August 2, 1854; Series 5, vol. 1, item 125, RG 49, NARA-Seattle. Ives listed T. 36, 37 & 38 S., R. 1 E; T. 35, 36, 37 & 38 S,. R 1 W.; T 36 & 37 S., R 2 W, & T 36 S., R 3W, adding "The subdividing in the remaining ten townships will all be fractional & about the amount that it is thought would be necessary to subdivide."

[22] Hyde selected Township 36 South, Range 1 East, and Ives Township 38 South, Range 1 East.

[23] BLM Field Notes, vol. OR-R0033 p. 0490.0, Township 36 South, Range 2 West (Subdivision).

[24] Township 36 South, Range 2 West, Original Field Notes of Ives and Hyde, vol. 19, 40–41.

[25] BLM Field Notes, vol. OR-R0033, p. 0490.0, Township 36 South, Range 2 West (Subdivision).

[26] Contract No. 47; Series 22; RG 49 SG/CS; NARA-Seattle.

[27] Gardner to Wilson, July 12, 1854; Series 2, vol. A., p. 238; Series 2, RG 49, NARA-Seattle.

[28] Ibid., 247.

[29] Gardner's action adjusted the contract payment originally outlined at $25 per mile for the survey of the Willamette Meridian, $17 per mile for the Eighth Standard Parallel, and $11 per mile for the survey of exterior lines for twenty townships and for the subdivision of ten townships; Gardner to Wilson, Series 2, vol. A, pp. 257–260, RG 49, NARA-Seattle.

[30] Contract No., 47; Series 22; RG 49 SG/CS; NARA-Seattle. Their relinquishment stipulated, "All other provisions and conditions of the contract including the Twenty-five dollars per mile for Meridian and meridian offset lines to remain in full force and binding on each party to the said contract."

[31] McFall, September 9–10, 1854.

[32] Orson Avery Stearns, *Reminiscences of Pioneer Days and Early Settlers of Phoenix and Vicinity* (Ashland, Oregon, 1922).

[33] BLM Field Notes, vol. OR-R0039, pp. 0384.0–0385.0; McFall, October 11, 1854. The surveyors worked in Township 35 South, Range 2 West.

[34] BLM Field Notes, vol. OR-R0039, p. 0671.0, Township 38 South, Range 2 West (Subdivision). Hyde next subdivided along the North and South Forks of Little Butte Creek and, from September 28 to October 7, worked in Township 35 South, Range 1 East, its settled areas lying along Little Butte Creek and on Antelope Creek.

[35] Hyde worked in Township 34 South, Range 1 East, and Township 35 South, Range 2 West, Original Field Books of Ives and Hyde, Book 9, p. 60.

[36] McFall, October 22, 1854.

[37] Patterson, a physician and a surveyor, settled in Lane County in 1852 and served in the Territorial Legislature from 1853 to 1854.

[38] Contract No. 45 dated July 10, 1854, was the last issued by the Oregon surveyor general for surveys north of the Columbia River. It called for the survey of the Third and Fourth Standard Parallels North through Range 3 West; exterior lines for Townships 13 through 16 North, Range 3 West, subdivisions for Townships 14 through 16 North, Range 3 West, and the survey of the First Standard Parallel North, through Townships 2 and 3 West.

[39] In 1854, the claims contractors were Alleck Smith, John P. Welsh, Luther White, Isaac R. Moore, Addison R. Flint, Walter Howard, Dennis Cathorn, Luther D. Kennedy, Nelson Simons, Leonard B. Vickers, Ebenezer Haft, J. Addison Pownall, W. Patterson, David Stump, William Logan, Abraham Sulger, Thaddeus Hamson, V. W. Crawford, A. S. Watt, Ezra Fisher, Reuben Ford, and Elliott Bowman.

[40] Gardner to Wilson, August 1, 1854; Series 2, vol. A., p. 238, RG 49, NARA-Seattle.

[41] Alexander Gordon to Charles Gardner, August 7, 1854, Series 3, vol. 5, p. 57, RG 49, NARA-Seattle.

[42] Amanda Hardin to Charles Gardner, September 4, 1854; Series 3, vol. 5, p. 65, RG 49, NARA-Seattle.

[43] Ibid.; Charles Gardner to William G. T'Vault, September 13, 1854; Series 3, vol. 5, p. 66, RG 49, NARA-Seattle.

[44] Gardner to Wilson, October 9, 1854; Series 2, vol. A, p. 246, RG 49, NARA-Seattle.

[45] Ibid., November, 21, 1854, pp. 252–253.

[46] Beckham, *Land of the Umpqua*, 162.

[47] White, "The Public Land Surveys in Oregon." The Oregon City Land Office served the Willamette Valley, and the office at Winchester helped settlers in southwestern Oregon. A third land office was planned for Olympia in Washington Territory. President Pierce appointed James Tilton as Washington Surveyor General in August, 1854. Tilton, however, would not arrive until the following spring.

[48] National Archives and Record Service, Pamphlet Describing M815, "Oregon and Washington Donation Land Files 1851–1903, National Archives and Records Service, Washington, DC, 1973, 2; *Oregon Spectator,* December 5, 1854, 3:1. The provision for the district land office was made by Act of Congress, approved July 17, 1854, amending donation land laws.

[49] *Oregon Spectator*, December 2, 1854, 2:5. The announcement was dated November 10, 1854. In addition, all conflicts regarding donation land claims, not adjudicated before December 11, 1854, would be transferred to the Register and Receiver for resolution.

50 Gardner to Wilson, December 20, 1854; Series 2, vol. A, p. 269, RG 49, NARA-Seattle. James Tilton to Charles Gardner, September 5, 1854 and September 15, 1854; Series 5, vol. 1, items 130; 161, RG 49, NARA-Seattle; White, *A History of the Rectangular Survey System*, 222. Tilton arrived in Olympia in March, 1855.

51 Gardner to Hendricks, December 17, 1855; Series 2, vol. A., pp. 332–333, RG 49, NARA-Seattle.

52 *Oregon Statesman*, December 12, 1854, 2:3.

53 *Oregon Statesman*, December 12, 1854, 2:4; June 13, 1854, 2:2; September 19, 1854, 2:7.

Chapter 11

1 *Oregon Statesman*, January 23, 1855, 3:3; N. de Bertrand Lugrin, *The Pioneer Women of Vancouver Island 1843–1866,* 306–307; Hollis R. Lynch, "Sir Joseph William Trutch, a British-American Pioneer on the Pacific Coast," *Pacific Historical Review* 30, no. 3 (1961): 248. Julia and Joseph Trutch moved to Illinois in 1856 where he became an assistant superintendent for the Illinois and Michigan Canal. In 1859, the couple settled in British Columbia. Trutch became British Columbia's first lieutenant governor.

2 *Oregon Statesman*, December 26, 1854 3:3 (advertisement). In later years, George Hyde told his son that Preston's and O'Neill's business had not succeeded as they had hoped.

3 *Oregon Statesman,* December 19, 1854, 3:1; Oregon *Spectator*, August 11, 1854, 2:2; January 13, 1855, 1:1–5.

4 *Oregon Statesman*, June 20, 1854 2:4; Whig candidate Preston received twenty-three more votes than his opponent, Democrat Philip Foster.

5 *Oregon Spectator*, January 6, 1855 2:4.

6 *Oregon Statesman*, February 10, 1855 2:2. Preston published the map, Preston's "Sectional and County map of Oregon and Washington West of the Cascade Mountains," (Chicago, Illinois: A. H. Burley, 1856), the next year and offered it for sale.

7 Michael Edmonds, "The U.S. General Land Office and Commercial Map Making: A Case Study," *Government Publications Review* 13 (1986): 571.

8 *Oregon Spectator*, February 17, 1855, 2:2; *Oregon Statesman,* February 20, 1855, 2:5 (advertisement); *Oregon Statesman* December 26, 1854, to March 27, 1855 (advertisement).

9 *Oregon Statesman*, February 27, 1855, 2:7; Johnson, *Founding the Far West*, 43. With the expiration of the Donation Land Act in 1855, the number of emigrants dropped sharply to five hundred that year.

10 Gardner to Wilson, October 23, 1854, Series 2, vol. A, p. 247, RG 49, NARA-Seattle.

11 White, *A History of the Rectangular Survey System,* 119; William Burt to Charles Gardner, May 15, 1854, Series 5, Box 120, vol. 1, p. 118, RG 49, NARA-Seattle.

12 The Territorial Legislature created Josephine County on January 22, 1856, from the western half of Jackson County.

13 "Addenda to the Usual Special Instructions, February 19, 1855." The contract specified the following for survey: "Townships 31 & 32 S. 5 and 6 W. 33 & 34 S. 6 W; 35 & 36 S. 6 and 7 W; 7th Standard Parallel south through range 6 W; 8th Standard Parallel south through ranges 6 and 7 W, thence by offsets to Illinois Valley and the township and subdivision lines of ten townships in Illinois Valley."

14 Lake to Gardner, March 26, 1855, Series 5, vol. 1, item 170, RG 49, NARA-Seattle.

15 BLM Field Notes, vol. OR-R0044, p. 0218.0.

16 Hyde and his men first subdivided Township 33 South, Range 6 West.

17 Robbins, *Landscapes of Promise*, 85; Stephen Dow Beckham, *Requiem for a People: The Rogue Indians and the Frontiersmen* (Norman, Oklahoma: University of Oklahoma Press, 1971), 149.

18 Lake to Gardner, June 19, 1855; Series 5, vol. 1, p. 305, RG 49, NARA-Seattle.

19 Lake and Hyde decided which townships to survey and subdivide as they progressed through the region.

20 BLM Field Notes, vol. OR-R0045, p. 0606.0; vol. OR-R0045, p. 0391.B, and vol. OR-R0045, p. 0540.0.

21 Waldo, Oregon, is located in the upper Illinois Valley in Township 40 South, Range 8 West.

22 BLM Field Notes, vol. OR R0045, p. 0607.0; vol. OR-R0045, p. 0159.0, and vol. OR-R0045, p. 0159B.

23 Gardner awarded these contracts to Matthew Murphy, T. H. Hutchinson, John P. Welsh, Ebenezer H. Hadt, Luther O. Kennedy, Ezra Fisher, Addison Powell, Sewall Truax, Elliott Bowman, J. A. Burnett, and Samuel D. Snowden.

24 Robert Metcalf to Charles Gardner, February 19, 1855; Series 5, vol. 1, p. 144, RG 49, NARA-Seattle.

25 Preston to Gardner, November 16, 1854; Series 5, vol. 1, p. 169, RG 49, NARA-Seattle; Lake to Gardner, March 26, 1855; Series 5, vol. 1, p. 172, RG 49, NARA-Seattle.

26 "A Citazen [sic]of Rogue River" to Charles Gardner, August, 15, 1856; Series 3, vol. 5, RG 49, NARA-Seattle.

27 Gardner awarded Truax Contract 63 on September 6, 1855.

28 White, *A History of the Rectangular Survey System*, 112, 194. Hendricks was appointed Commissioner on August 8, 1855.

29 Charles Gardner, Surveyor General to Thomas Hendricks, Commissioner, General Land Office, December 17, 1855; Series 2, vol. A, pp. 331–334, RG 49, NARA-Seattle.

30 *Oregon Statesman,* May 6, 1856, 2:2–4; May 27, 1856, 2:3; Oregon State Archives, *Guide to Oregon Provisional and Territorial Government Records*, Oregon State Archives, 1990, p. 80. The 7th Session of the Legislature met from December 3, 1855, to January 31, 1856.

31 Ibid., May 6, 1856, 2: 2–4.

32 Ibid.

33 Ibid., May 27, 1856, 2:3–4.

34 Ibid.

35 White, *A History of the Rectangular Survey System*, 219; John Zieber replaced Gardner as Surveyor General on March 18, 1856. Sixty-nine years old when he left Oregon, Gardner served as Clerk of the US Treasury from 1857 to 1867. He died in Washington, DC, on November 1, 1869, and is buried in the Congressional Cemetery.

36 Stephen Dow Beckham, "Oregon History — Overland to Oregon," *Oregon Blue Book*, http://www.bluebook.state.or.us/cultural/history/history12.htm (May 15, 2004). Ultimately, 7,437 settlers, successfully acquired land claims in Oregon.

37 Prosch, "Notes on Oregon Conditions in the Fifties," 198.

38 John Zieber, Surveyor General to Thomas Hendricks, Commissioner General Land Office, June 17, 1856; Series 2, vol. A, pp. 356–357, RG 49, NARA-Seattle. Only five surveyors would take contracts in 1856: Joseph Trutch, Dennis Hathorn, A. W. Patterson, Alleck C. Smith, and Samuel Snowden.

[39] Malcolm Clark, Jr., *Eden Seekers, The Settlement of the Oregon Country, 1818–1862* (Boston, Massachusetts: Houghton Mifflin Company, 1981), 277; Lang, *History of the Willamette Valley* vol. 1, 391–392; Provisional and Territorial Documents, Microfilm Reel 58, Document 6504, Oregon State Archives, Salem, Oregon.

[40] Preston to Hyde, July 14, 1855, Louis Hyde Family Book, Reed Hyde Papers.

[41] White, "The Public Land Surveys in Oregon."

[42] Stuart Allan, Aileen R. Buckley, and James E. Meacham. *Atlas of Oregon,* edited by William G. Loy. (Eugene, Oregon: University of Oregon Press, 2001), 19.

[43] Jeff LaLande, personal communication. August 12, 2004.

[44] Stuart Allan et al., 27–28.

[45] Johansen and Gates, *Empire of the Columbia*, 236.

[46] Stephen Dow Beckham, *Cultural Resource Overview of the Siskiyou National Forest*, Grants Pass, Oregon: USDA Forest Service, Siskiyou National Forest, 1978, 71.

[47] Dorothy O. Johansen, "The Roll of Land Laws in the Settlement of Oregon," *Genealogical Material in Oregon Donation Land Claims*, vol. 1 (Portland, Oregon: Genealogical Forum of Portland, 1957).

[48] Beckham, *Land of the Umpqua*, 163.

[49] Beckham, "Oregon History — Territorial Government" *Oregon Blue Book*, http://bluebook.state.or.us/cultural/history/history12.htm (May 15, 2004).

[50] Bowen, *The Willamette Valley: Migration and Settlement on the Oregon Frontier*, 103. The surveyors accurately noted the points where natural or man-made features intersected their lines. Draftsmen in the surveyor general's office could not always precisely plot the extent of these features between the mathematically calculated positions.

Epilogue

[1] William Ives to Butler Ives, May 8, 1864, quoted in Isabella E. Swan, *The Deep Roots: A History of Grosse Ile, Michigan to July 6, 1876* (Grosse Ile, Michigan: Isabella Swan, 1976), 294–295; Silas Farmer, *History of Detroit and Wayne County and Early Michigan* (Detroit, Michigan: Silas Farmer & Co., 1890; reprinted by Gale Research Co., Detroit Michigan, 1969), 130. In 1996, the William Ives Trail at Lacey, Washington, was dedicated through the efforts of the Land Surveyors' Association of Washington Historical Society.

[2] White, "Willamette Initial Point History," *The Oregon Surveyor* 21, no. 3 (1993): 20; Bureau of Land Management, General Land Office Records, http://www.glorecords.blm.gov/PatentSearch/.

[3] Virginia City, Nevada, *Territorial Enterprise*, June 6, 1866, 3; Henry G. Langley (compiler), *San Francisco Directory for Year Commencing December 1865* (San Francisco, California: Henry G. Langley, 1865); L. M. McKenney (compiler), *Pacific Coast Directory 1886–1887* (San Francisco, California: L. M. McKenney & Co., 1886), 182: Francois D. Uzes, *Chaining the Land: A History of Surveying in California* (Sacramento, California: Landmark Enterprises, 1977), 290; *San Francisco Morning Call*, June 26, 1895, 13:7; *San Francisco Chronicle,* June 25, 1895, 12:4.

[4] White, *A History of the Rectangular Survey System*, 130; Landrum, "A Major Monument: Oregon — California Boundary," 14.

[5] Butler Ives to William Ives, April 2, 1866; July 19, 1868, Timothy Hopkins Transportation Collection.

[6] *Sacramento Daily Union*, December 29, 1971, 3:5.

[7] George G. Clark, Wells Fargo and Company Express, to A. Ives & Son, Detroit Michigan, January 4, 1872, Timothy Hopkins Transportation Collection.

[8] *History of Will County Illinois*, 403; Louis Hyde Family Book, Reed Hyde Papers.

[9] Wells Lake to John Zieber, March 19, 1858; Series 3, vol. 7; RG-49; NARA-Seattle. Lake left Oregon after 1855 to return east. He married Irene Noxon Trowbridge at Syracuse, New York, on October 25, 1860, and, by 1870, the couple lived at Evanston, Illinois, where Lake worked as a commission merchant.

[10] *Joliet Republic and Sun*, December 4, 1891, p. 3; *Joliet Daily News* November 30, 1891, 3.

[11] *Oregon Statesman*, June 7, 1859, 2:7.

[12] *Joliet Signal,* April 18, 1865: 2; *History of Will County Illinois,* 307; *Portland Oregonian*, May 26, 1865, 2:1.

[13] Charles Davies, *Elements of Surveying and Navigation*, 127–128.

Selected Bibliography

Manuscripts

Chadwick, Stephen Fowler. "Oregon Party." Yale Collection of Western Americana, The Beinecke Rare Book and Manuscript Library, Manuscript WA MSS 71. Yale University, New Haven, Connecticut.

Hyde, Louis. Reed Hyde Papers, Private Collection.

Ives, Butler. Diaries 1851–1853, 1855, 1860. Western Reserve Historical Society, Manuscript 3011, container 1, Folders 1, 2, and 3. Western Reserve Historical Society, Cleveland, Ohio.

———. Surveying Contracts and Appointments, 1850–1861, Timothy Hopkins Transportation Collection, (MO97, Box 6, Folder 1), Department of Special Collections and University Archives, Stanford University Libraries, Stanford, California.

Ives, William. Correspondence and Papers, 1847–1864. Burton Historical Collection, Manuscript 1126. Burton Historical Collection, Detroit Public Library, Detroit, Michigan.

Ives, William. Diaries. Western Reserve Historical Society, Manuscript 3011, Container 1, Folder 2. Western Reserve Historical Society, Cleveland, Ohio.

McFall, George. Journal. Typescript held in the office of the Jackson County Surveyor, Medford, Oregon.

Nelson, Thomas. Letters from Oregon 1851–1853. The Yale Collection of Western Americana, The Beinecke Rare Book and Manuscript Library, Manuscript WA MSS S-2176. Yale University, New Haven, Connecticut.

Preston, John B. Letters from Donation Land Claimants, 1853. Oregon Historical Society Research Library, Manuscript 914. Oregon Historical Society, Portland, Oregon.

Preston, John B. John B. Preston Collection. Oregon Historical Society Research Library Manuscript 914. Oregon Historical Society, Portland, Oregon.

Books

Allan, Stuart, Aileen R. Buckley, and James E. Meacham. *Atlas of Oregon.* 2nd edition. Edited by William G. Loy. Eugene, Oregon: University of Oregon Press, 2001.

Atwood, Katherine C., and Frank A. Lang. *As Long as the World Goes On: Environmental History of the Evans Creek Watershed.* USDI Bureau of Land Management, Medford District Office, Oregon, 1995.

Bancroft, Hubert Howe. *History of Oregon, 1848–1888.* 2 volumes. San Francisco, California: The History Company, 1888.

Beckham, Stephen Dow. *Requiem for a People: The Rogue Indians and the Frontiersmen.* Norman, Oklahoma: University of Oklahoma Press, 1971.

———. *Land of the Umpqua: A History of Douglas County, Oregon.* Roseburg, Oregon: Douglas County Commissioners, 1986.

Boag, Peter G. *Environment and Experience: Settlement Culture in Nineteenth Century Oregon.* Berkeley, California: University of California Press, 1992.

Bowen, William A. *The Willamette Valley: Migration and Settlement on the Oregon Frontier.* Seattle, Washington: University of Washington Press, 1978.

Burt, William A. *A Key to the Solar Compass, and Surveyor's Companion.* Philadelphia, Pennsylvania: William S. Young, 1855.

Caldwell, Norman C. *Surveyors of the Public Lands in Michigan 1808–2000.* Owosso, Michigan: Norman Caldwell, 2001.

Cazier, Lola. *Surveys and Surveyors of the Public Domain 1785–1975.* Washington, DC: US Department of the Interior, 1977.

Clark, Malcolm, Jr. *Eden Seekers: The Settlement of the Oregon Country, 1818–1862.* Boston, Massachusetts: Houghton Mifflin Company, 1981.

Corning, Howard McKinley. *Willamette Landings: Ghost Towns of the River.* Portland, Oregon: Binfords & Mort, 1947.

———. *Dictionary of Oregon History.* Portland, Oregon: Binfords & Mort, 1956.

Davies, Charles. *Elements of Surveying and Navigation: With a Description of the Instruments and the Necessary Tables.* New York, New York: A. S. Barnes & Co., 1847.

———. *Elements of Surveying and Navigation: With a Description of the Instruments and the Necessary Tables.* New York, New York: A. S. Barnes & Co. 1854.

Duniway, David C. *Members of the Legislature of the State of Oregon, 1843–1967.* Salem, Oregon: Oregon State Archives, 1968.

Farmer, Silas. *History of Detroit and Wayne County and Early Michigan.* Detroit, Michigan: Silas Farmer & Co. 1890. (Reprint) Detroit, Michigan: Gale Research Company, 1969.

Ficken, Robert E. *Washington Territory.* Pullman, Washington: Washington State University Press, 2002.

Franklin, Jerry F., and C. T. Dyrness. *Natural Vegetation of Oregon and Washington.* Corvallis, Oregon: Oregon State University Press, 1988.

Genealogical Material in Oregon Donation Land Claims. Volume I. Portland, Oregon: Genealogical Forum of Portland, 1957.

Genealogical Material in Oregon Donation Land Claims Volume II. Portland, Oregon: Genealogical Forum of Portland, Oregon, 1959.

Genealogical Material in Oregon Donation Land Claims Volume III. Portland, Oregon: Genealogical Forum of Portland, Oregon, 1962.

Gurley, Lottie L. *Genealogical Material in Oregon Donation Land Claims.* Volume V. Portland, Oregon: Genealogical Forum of Portland, Oregon, 1975.

Gurley, W., & L. E. Gurley. *A Manual of the Principal Instruments Used in American Engineering and Surveying.* Troy, New York: W. & L. E. Gurley, 1874. (Reprint): Mendham, New Jersey: The Astragal Press, 1993.

————. *A Manual of the Principal Instruments Used in American Engineering and Surveying.* Troy, New York: W. & L. E. Gurley, 1910.

Heyl, Erik. *Early American Steamers.* Volumes 1–6. Buffalo, New York: Erik Heyl, 1969.

History of Will County, Illinois. Chicago, Illinois: Wm. Le Baron Jr. & Co., 1878.

Hussey, John A. *Champoeg: Place of Transition, A Disputed History.* Portland, Oregon: Oregon Historical Society Press, 1967.

Johansen, Dorothy O., and Charles M. Gates. *Empire of the Columbia: A History of the Pacific Northwest.* New York, New York: Harper and Row, 1967.

Johnson, David Alan. *Founding the Far West: California, Oregon and Nevada, 1840–1890.* Berkeley, California: University of California Press, 1992.

LaLande, Jeff. *An Environmental History of the Little Applegate River Watershed, Jackson County, Oregon.* USDA Forest Service, Rogue River National Forest, 1995.

Lang, Herbert O. *History of the Willamette Valley.* Portland, Oregon: Himes and Lang, 1885.

Linklater, Andro. *Measuring America: How an Untamed Wilderness Shaped the United States and Fulfilled the Promise of Democracy.* New York, New York: Walker and Co., 2002.

Loy, William G., editor. *Atlas of Oregon.* Eugene, Oregon: University of Oregon Press, 1976.

Lugrin, N. de Bertrand. *The Pioneer Women of Vancouver Island 1843–1866.* Victoria, British Columbia: The Women's Canadian Club of Victoria, 1928.

Meany, Edward S. *Origin of Washington Geographic Names*. Seattle, Washington: University of Washington Press, 1923.

Minnick, Roy. "Instructions to Surveyor of Public Lands in Oregon, 1851." *A Collection of Original Instructions to Surveyors of the Public Lands 1815–1881*. Rancho Cordova, California: Landmark Enterprises, 1992.

Minter, Harold Avery. *Umpqua Valley Oregon and Its Pioneers*. Portland, Oregon: Binfords & Mort, 1967.

Moore, John M., compiler. *Instructions to the Surveyors' General of Public Lands for the Surveying Districts Established in and Since the Year 1850 Containing Also A Manual of Instructions*. Washington, DC: A. O. P. Nicholson, Public Printer, 1855.

Rasmussen, Louis J. *San Francisco Ship Passenger Lists*. Volume 2. Baltimore, Maryland: Deford & Co., 1966.

Robbins, William G. *Landscapes of Promise: The Oregon Story 1800–1940*. Seattle, Washington: University of Washington Press, 1997.

Schwantes, Carlos Arnaldo. *Long Day's Journey: The Steamboat & Stagecoach Era in the Northern West*. Seattle, Washington: University of Washington Press, 1999.

Snowden, Clinton. *A History of Washington State: The Rise and Progress of an American State*, Volume III. New York, New York: The History Co., 1909.

Stewart, Lowell O. *Public Land Surveys*. New York, New York: Arno Press, 1979.

Uzes, Francois D. *Chaining the Land: A History of Surveying In California*. Sacramento, California: Landmark Enterprises, 1977.

———. *Illustrated Price Guide to Antique Surveying Instruments and Books*. Rancho Cordova, California: Landmark Enterprises, 1980.

Walling A. G. *History of Southern Oregon: Comprising Jackson, Josephine, Douglas, Curry and Coos Counties*. Portland, Oregon: A. G. Walling, 1884.

Webster, Kimball. *The Gold Seekers of '49: A Personal Narrative of the Overland Trail and Adventures in California and Oregon from 1849 to 1854*. Manchester, New Hampshire: Standard Book Company, 1917.

White, C. Albert, *A History of the Rectangular Survey System*. Washington, DC: US Department of the Interior, Bureau of Land Management, 1983.

———. *Initial Points of the Rectangular Survey System*. Westminster, Colorado: Publishing House/Professional Land Surveyors of Colorado, 1996.

———. *A Casebook of Oregon Donation Land Claims*. Oregon City, Oregon: LLM Publications, 2001.

Woodward, Walter Carleton. *The Rise and Early History of Political Parties in Oregon 1843–1868*. Portland, Oregon: J. K. Gill Company, 1913.

Articles

Barker, Burt Brown. "The Estate of Dr. John McLoughlin: The Papers Discovered." *Oregon Historical Quarterly* 50 (1949): 155–185.

Bentson, William Allen. "Historic Capitols of Oregon: An Illustrated Chronology." Salem, Oregon: Oregon Library Foundation. http://www.osl.state.or.us/lib/capitols/index.htm.

Bouman, Lane J. "The Location and Survey of Oregon Donation Land Claims," Part 1, "The Role of the Surveyor in the Initial Location of Donation Land Claims." *The Oregon Surveyor* 32 (August–September, 1999): 14–15.

―――. "The Location and Survey of Oregon Donation Land Claims," Part 2, "The Role of the Surveyor in Final Monumentation of Donation Land Claims." *The Oregon Surveyor* 32 (November–December,1999): 16–18.

Burt, John S. "The Search for an Ancestor: Pioneer Surveyor William A. Burt." *The California Surveyor* (Summer 1981): 10–13.

"Cost of Improvements Made [by Dr. John McLoughlin] at Willamette Falls to January 1, 1851." *Oregon Historical Quarterly* 14 (1913): 68–70.

Cutting, Maria Cable. "After Thoughts." *Oregon Historical Quarterly* 63 (1962): 237–241.

Dierking, Eric. "Survey Contracts for Surveying of Oregon Territory," Seattle, Washington: National Archives and Records Administration-Pacific Alaska Region, 1983.

Foote, Francis Seeley. "The Boundary Line between California and Oregon." *California Historical Society Quarterly* 19 (1940): 368–372.

Gilchrist, Marie E. "A Michigan Surveyor — William Ives" *Inland Seas* 21 (1965): 312–321.

―――. "Isle Royale Survey," part 1, *Inland Seas* 24(1968): 179–185.

―――. "Isle Royale Survey," part 2 *Inland Seas* 24 (1968): 303–4, 313–25.

―――. "Isle Royale Survey," part 3. *Inland Seas* 35 (1969): 26–40.

Gilchrist, Marie E., editor. "William Ives' Huron Mountains Survey, 1846." *Michigan History* 50 (1966): 323–340.

Glenn, William W., and Jeanne E Glenn. "Public Lands Surveyors in the Northwest." *The Oregon Surveyor* 18 (March–April, 1990): 27–28.

―――. "Early Public Land Surveyors in the Northwest." *The Oregon Surveyor* 18 (July–August, 1990): 21–22.

―――. "Early Public Lands Surveyors in the Northwest." *The Oregon Surveyor* 19 (January–February, 1991): 30–31.

Johansen, Dorothy, O. "The Roll of Land Laws in the Settlement of Oregon." *Genealogical Material in Oregon Donation Land Claims*. Volume I, Portland, Oregon: Genealogical Forum of Portland, 1957.

LaLande, Jeff. "'Dixie' of the Pacific Northwest: Southern Oregon's Civil War." *Oregon Historical Quarterly* 100 (1999): 32–81.

Landrum, Francis. "A Major Monument: Oregon-California Boundary." *Oregon Historical Quarterly* 72 (1971): 5–53.

Lynch, Hollis R. "Sir Joseph William Trutch, a British-American Pioneer on the Pacific Coast." *Pacific Historical Review* 30 (1961): 243–256.

McNeil, F. H. "Official Survey of Oregon Country Was Undertaken 65 Years Ago Today." *Oregon Sunday Journal,* June 4, 1916, Sec. 2:9.

Perry, James R., Richard H. Chused, and Mary De Lano, editors. "The Spousal Letters of Samuel Royal Thurston, Oregon's First Territorial Delegate to Congress 1849–1851." *Oregon Historical Quarterly* 96 (1995): 4–79.

Prosch, Thomas W. "Notes on Oregon Conditions in the Fifties." *Oregon Historical Quarterly* 8 (1907): 191–200.

Rockwood, E. Ruth, editor. "Diary of G. H. Atkinson, 1847–1858." *Oregon Historical Quarterly* 40 (1939): 275–282.

———. "Diary of G. H. Atkinson, 1847–1858." *Oregon Historical Quarterly* 41 (1940): 212–226, 288–303.

Spickelmier, Verda. "The Oregon Territory's First Surveyor General Left His Mark." Oregon City *Enterprise Courier,* 15 December, 1991, Section 3A.

Squires, Rod. "The Public Land Survey in Minnesota Territory, 1847–1852," http://www.geog.umn.edu/faculty/squires/.

Stegner, Wallace. "History Comes to the Plains" *American Heritage* 8 (June 1957): 14–19+.

Toscano, Patrick. "The Gunter's Chain" *Surveying and Land Information Systems* 51 (1991): 155–161.

US General Land Office. Series 2, Letters Sent to Commissioner, General Land Office 1851–1859. Record Group 49, Records of the Bureau of Land Management. National Archives and Records Administration-Pacific Alaska Region.

———. Series 3, Miscellaneous Letters Received 1851–1913. Record Group 49, Records of the Bureau of Land Management. National Archives and Records Administration.

———. Series 4, Miscellaneous Letters Sent, 1851–1921, Record Group 49. Records of the Bureau of Land Management. National Archives and Records Administration.

———. Series 5, Letters Received from Deputy Surveyors 1851–1902. Record Group 49. Records of the Bureau of Land Management. National Archives and Records Administration-Pacific Alaska Region.

————.Series 22, Contracts and Bonds 1851–1870. Record Group 49. Records of the Bureau of Land Management, (Records of the Surveyor General of Oregon and the Oregon Cadastral Survey Office). National Archives and Records Administration-Pacific Alaska Region.

————. Series 25, Copies of Field Notes of Contract Surveys. Record Group 49. Records of the Bureau of Land Management, (Records of the Surveyor General of Oregon and the Oregon Cadastral Survey Office). National Archives and Records Administration-Pacific Alaska Region.

————. Series 84, Township Plats 1852–1906. Record Group 49. Records of the Bureau of Land Management, National Archives and Records Administration-Pacific Alaska Region.

Vaughan, Thomas, editor. "The Round Hand of George B. Roberts." *Oregon Historical Quarterly* 63 (1962): 101–236.

Warner, Deborah Jean. "William J. Young. From Craft to Industry in a Skilled Trade," *Pennsylvania History* 52 (1985): 53–68.

Weilepp, Bruce. "Slow Boats & Fast Water: A Maritime History of the Cowlitz River." *Columbia* 10 (1996): 16–22.

White, C. Albert. "Willamette Initial Point History," Part 1. *The Oregon Surveyor* 21 (May–June, 1993): 14–16.

————. "Willamette Initial Point History," Part 2. *The Oregon Surveyor* 21 (July–August, 1993): 8–21.

————. "Willamette Initial Point History," Part 3. *The Oregon Surveyor* 22, (February–March, 1994): 18–21.

————."Why Boustrophedonic?" *The Oregon Surveyor* 34 (June–July, 2001): 19–23.

————. "Harvey Gordon, U.S. Deputy Surveyor and Designer of the Oregon State Seal." *The Oregon Surveyor* 35 (April–May, 2002): 18–22.

————. "The Public Land Surveys in Oregon." Professional Land Surveyors of Oregon, http://www.plso.org/readingroom/GLO.htm (January 3, 2004).

"Willamette Stone Rededication and 150th Year Commemoration Celebration, Saturday, June 19, 2001." *The Oregon Surveyor* 34(4; August–September, 2001): 14–15.

Winthur, Oscar Osburn. "Inland Transportation and Communication in Washington 1844–1859." *Pacific Northwest Quarterly* 30 (1939): 371–386.

Yonce, Fred. "Public Land Surveys in Washington." *Pacific Northwest Quarterly* 63 (1972):129–141.

URLs

USDI Bureau of Land Management General Land Office Records: http://www.glorecords.blm.gov/.

US Department of the Interior Land Status & Cadastral Records Viewer, Willamette Meridian — Oregon and Washington States (BLM Field Notes): http://www.or.blm.gov/or/landrecords/landrecords.php (January 30, 2008).

University of Oregon Libraries, Map & Aerial Photography Collection General Land Office Quadrant Reference Page: http://libweb.uoregon.edu/map/GIS/Data/.

Index

257

258

263